NANOMATERIALS AND NANO-BIOCHAR IN REDUCING SOIL STRESS

An Integrated Approach to Sustainable Agriculture

NANOMATERIALS AND NANO-BIOCHAR IN REDUCING SOIL STRESS

An Integrated Approach to Sustainable Agriculture

Edited by

Vishnu D. Rajput, PhD

Tatiana M. Minkina, PhD

AAP | APPLE ACADEMIC PRESS

First edition published 2025

Apple Academic Press Inc.
1265 Goldenrod Circle, NE,
Palm Bay, FL 32905 USA

760 Laurentian Drive, Unit 19,
Burlington, ON L7N 0A4, CANADA

CRC Press
2385 NW Executive Center Drive,
Suite 320, Boca Raton FL 33431

4 Park Square, Milton Park,
Abingdon, Oxon, OX14 4RN UK

Library and Archives Canada Cataloguing in Publication

CIP data on file with Canada Library and Archives

Library of Congress Cataloging-in-Publication Data

CIP data on file with US Library of Congress

ISBN: 978-1-77491-707-7 (hbk)
ISBN: 978-1-77491-831-9 (pbk)
ISBN: 978-1-00350-345-3 (ebk)

About the Editors

Vishnu D. Rajput, PhD
Principal Scientist, Academy of Biology and Biotechnology, Southern Federal University

Vishnu D. Rajput, PhD, is working as a Lead Researcher (Associate Professor) at Southern Federal University, Rostov-on-Don, Russia. His ongoing research is based on the toxic effects of bulk- and nano-forms of metals and the bioaccumulation; bio/geo-transformations; uptake, translocation, and toxic effects on plant physiology; morphology, anatomy, ultrastructure of cellular and subcellular organelles; cytomorphometric modifications; and DNA. He is also researching possible remediation approaches such using biochar/nano-biochar-based sorbents and nanotechnology. With his vast experience and experimental work, he has comprehensively detailed the state of research in environmental science regarding how nanoparticles/heavy metals interact with plants, soil, microbial community, and the larger environment as well as possible remediation technology using nanoparticles/nano-biochar. Dr. Rajput has published over 412 scientific publications, including nearly 296 peer-reviewed full-length articles, 16 books (Springer, Elsevier, CRC Press, Nova-USA), and many chapters (Scopus-indexed) and conference articles. His h-index is 41 by Scopus with 6,786 citations as of July 2024. He is an internationally recognized reviewer, having reviewed over 250 manuscripts; he has received outstanding reviewer certificates from Elsevier, Springer, and *Plants* (MDPI). He is an editorial board member of several high-impact journals. He is a co-principal investor on several grants on the state of contaminated soils and plants using synchrotron methods and other related areas.

Tatiana M. Minkina, PhD
Head, Department of Soil Science and Land Evaluation, Southern Federal University

Tatiana M. Minkina, PhD, is the Head of Soil Science and Land Evaluation Department of Southern Federal University, Russia. She is Head of the International Master's Degree Educational Program "Management and Estimation of Land Resources" (2015–2022 years, accreditation by ACQUIN). Her area of scientific interest is soil science, biogeochemistry of trace elements, environmental soil chemistry, soil monitoring and assessment, and modeling and remediation using physicochemical treatment methods. Dr. Minkina was awarded in 2015 with a Diploma of the Ministry of Education and Science of the Russian Federation for her many years of work for the development and improvement of the educational process and for her significant contribution to the training of highly qualified specialists. Currently, she is handling projects funded by the Russian Scientific Foundation, Ministry of Education and Science of the Russian Federation, and the Russian Foundation of Basic Research. She is a member of many professional societies. She has published over 750 papers (389 in English).

Contents

Preface

Soil deterioration, increase in toxic elements in soil induced by rapid industrial development, and intensive cultivation impose a serious threat to global food security and environmental sustainability. Being a key living ecosystem, soil supports microorganisms, plants, as well as animals. In the soil system, numerous biological and biogeochemical activities occur. Improving these processes can help alleviate or adapt food production and agriculture to the changing climate and environment.

For the last few decades, soil fertility has been deteriorating rapidly and becoming a problem for achieving the zero-hunger goal. Enhancing soil productivity to improve soil fertility is the need of the hour, since sustainable food production is reliant on good soil health. Soil fertility and productivity depends on the interactions of microbes and other soil organisms. Nanotechnology presents a new hope in this direction.

The recent significant advances in nanotechnology have the capacity to expand current cultivation practices in a sustainable way. The use of nanomaterials can enhance microbial activities and the functionality of other soil organisms by applying or combining agricultural practices and soil health-improving amendments.

In this volume, the most promising carbon-rich materials, that is biochar and the more effective nano-biochar, are explored in various way. The chapters look at the production and characteristics in conjunction with the current status of nano-biochar, the benefits of biochar, and especially nano-enhanced biochar, are elaborated on. Considered in detail is the role of nano-biochar in metal stress reduction, in soil health improvement, nano-biochar as a soil conditioner, reducing soil stress using integrated approaches, improving nutrient use efficiency, salinity stress management, sustainable crop production, and arsenic remediation.

With its comprehensive coverage of the important topic of biochar and nano-biochar, the book will prove useful to companies, students, professors, researchers, and scientists who are interested in the topic of soil stress management and sustainable agriculture as well as policymakers who can recommend novel agriculture amendments.

CHAPTER 1

Nano- and Nano-Biochar: Overview, Production, and Characteristics

PRADEEP KUMAR,[1] KAJAL SINGH,[2] AMIT KUMAR SINGH,[3] NANCY SINGH,[4] SAKSHI SINGH,[5] SHRENI AGRAWAL,[6] RICHA DAS,[6] VISHNU D. RAJPUT,[7] TATIANA MINKINA,[7] SUNIL KUMAR MISHRA,[3] and KAVINDRA NATH TIWARI[1]

[1]*Department of Botany, MMV, Banaras Hindu University, Varanasi, Uttar Pradesh, India*

[2]*Department of Biosciences, Galgotias University, Greater Noida, Uttar Pradesh, India*

[3]*Department of Pharmaceutical Engineering and Technology, Indian Institute of Technology, Banaras Hindu University, Varanasi, Uttar Pradesh, India*

[4]*School of Basic and Applied Science, Galgotias University, Uttar Pradesh, India*

[5]*Department of Bioscience and Biotechnology, Banasthali Vidyapith, Niwai, Tonk, Rajasthan, India*

[6]*Department of Biotechnology, Parul Institute of Applied Science, Parul University, Vadodara, Gujarat, India*

[7]*Academy of Biology and Biotechnology, Southern Federal University, Rostov on Don, Russia*

Nanomaterials and Nano-Biochar in Reducing Soil Stress: An Integrated Approach to Sustainable Agriculture. Vishnu D. Rajput, & Tatiana Minkina (Eds.)

ABSTRACT

The global climate is shifting due to emissions of greenhouse gases and changes in agricultural practices. The atmosphere is being exposed to an increasing number of carbon-based molecules and pollutants as a result of human population growth, expanding factory output, and expanding agronomic practices. As a result, not only is wasted energy produced, but the atmosphere is also contaminated. With suitable techniques and management, these emissions could be converted to charcoal solids. Such carbon-based materials, often known as "biochar," are used in a wide range of industries, particularly agricultural production. Because of its crystalline characteristics, biochar can also be referred to as nano biochar (NBC). By using certain procedures, biochar can be transformed into NBC. In this case, the physiological and biological attributes of this valuable substance improve while better methods are used to reduce pollution discharges, remediate soil, produce energy, and manage waste. The planting area has recently expanded in an effort to increase crop yield. Scientists and farmers are now seeking ways to increase productivity per unit area as this technique has changed significantly. However, in other cases, increased production per unit area puts pressure on the cultivated land. If the situation is not properly managed, the soil will be overharvested of these essential nutrients, and eventually, the topsoil will be depleted of them. In certain circumstances, it is best to incorporate soil treatments. The soil is given various modifications to help with its physicochemical and biological features. Biochar is one of these materials that also improves the effectiveness of fertilizers and irrigation water. NBC enhances the conditions for crop growth and aids in effective soil conservation.

1.1 INTRODUCTION

Global effort is urgently required to tackle environmental contamination, which poses a threat to soil and water resources (Hou et al., 2020; WHO, 2015). According to Liu et al. (2022), contaminants can get into the atmosphere via both human and natural processes, such as industrial emissions, home sewage disposal, and agricultural practices. Contaminants are often divided into non-biodegradable and biodegradable categories (Gong et al., 2022; Samadi et al., 2021). Several pollutants like dyes, heavy metals, pharmaceutical waste products, and others are examples of non-biodegradable toxicants. Removal of these contaminants from the soil and water environments is a very tedious process that causes several abnormalities in living organisms. Almost 20 million hectares of agricultural land have already been contaminated by heavy metals worldwide (Liu et al., 2018). Long-term exposure to heavy metals has a severe impact on human health and can lead to a number of malignancies, hallucinations, sleep disorders, as well as other problems (Zhang et al., 2022). Despite the fact that organic pollutants degrade, certain remaining organic pollutants are frequently harmful and can cause malignancies in both people and animals, even at low concentrations (Liu et al., 2020; Singh et al., 2022). For instance, Khan et al. (2021) discovered that pharmaceutical ingredient remnants, which have even been reported in tiny amounts in surface water, might pose concern to the public's health in the aquatic environment. Many investigations have revealed that both aquatic and soil systems have collected extraordinary levels of pollution (Hou et al., 2020; Gong et al., 2022). Accomplishing soil and water restoration efforts is crucial.

The usage of biochars as efficient supplements for cleaning up contaminated water and soil has increased recently (Azadi & Raiesi, 2021; Chen et al., 2022; Luo et al., 2022; Wang et al., 2021). According to investigations (Li et al., 2020; Lian et al., 2020; Ren et al., 2022; Zhang et al., 2019), biochar may eliminate contaminants from polluted water and decrease contaminant accessibility in landscapes (Sun et al., 2022). Utilizing plant raw resources, biochar is pyrolyzed at various temperatures in the absence of oxygen environments (Li et al., 2022). As forestry and agricultural industries often provide the raw materials for the production of biochar wastes, it is affordable as well as ecologically sustainable, and it might also work well as a carbon-sequestering ingredient (Lyu et al., 2020; Liu et al., 2018). Nano biochar is defined as biochar with a particle diameter between 1 and 100 nm (Zhang et al., 2022). In general, there are two major approaches to produce nano biochar. One would be created artificially, whereas the second occurs when biochar physically degrades due to age in the environment (Huang et al., 2021; Ramanayaka et al., 2020). Several characteristic features like specific surface area (SSA), negative zeta potential, the number of oxygen-containing functional groups, as well as the hydrodynamic radius of nano biochar, are all significantly higher than those of bulk biochar (Ramanayaka et al., 2020; Lyu et al., 2018; Pratap et al., 2022).

NBC has been used in a wide range of applications, such as contaminant adsorbents, soil remedies, biosensing, and photocatalyst substances (Zhang et al., 2022; Dong et al., 2018; Xiao et al., 2020). The most reports revealed various utilization of nano biochar like cleaning up contaminated soils and water. According to studies by Lyu et al. (2018), and others, nano biochar can be a useful supplement for the removal of contaminants from contaminated water and stabilizing them in contaminated soil (Xiao et al., 2020; Zhang et al., 2021, 2022).

Nano biochar research is still considered a new area of research. Long-term studies on the uses of nano biochar in the environment are also insufficient. Despite the fact that there have been several publications summarizing relevant studies on nano biochar, these reviews either concentrate on the substance's characteristics or attempt to provide wide overviews of its use in a variety of sectors. To our knowledge, nano biochar's use in removing contamination from the environment has not yet been the subject of a thorough assessment. Consequently, the purpose of this chapter was to provide an overview of the NBC, biosynthesis, types, and its application in various sectors for the elimination of several contaminants from the environment (Figure 1.1).

1.2 BIOSYNTHESIS OF NANO-BIOCHAR

NBC is a new type of sophisticated nano powder that is currently being developed. At this time, the vast majority of NBC manufacturing and research is carried out using sustainable synthesis techniques and energy-saving approaches applicable to nanotechnology. Research conducted to date suggests that nano biochar is characterized by increased physical and chemical variations, such as increased surface areas, higher porosity, lower stokes radius, significant negative zeta potentials, structural features with high oxygen content, and carbon deformity that can also lead to the production of reactive organic species (ROS), while also increasing the capacity of adsorption, especially in contrast to its macrochar equivalent, and possessing exceptional features compared to black carbon (Zhang et al., 2021, 2022; Oleszczuk et al., 2016). This is in comparison to its macrochar counterparts, which have lower surface areas, higher porosity, and larger stokes radius.

Two methods, top-down and bottom-up, can be used to synthesize nanoparticles and create nanostructures (Lyu et al., 2018). The top-down strategy shrinks the solid materials to the minuscule level, whereas the bottom-up method is the opposite of the top-down method in which materials are made from the miniscule level. A few of the techniques used to create nanostructures are expensive. Consequently, a low-cost, environmentally friendly technique is needed to offset the high energy input, costly intermediates, and complex procedures that are often necessary in nanomaterials production methods (Qin et al., 2018). To save costs, several scientists have used mechanical techniques like cutting, etching, and grinding. Ball-milling is one of these techniques that is crucial for reducing the materials to the nanoscale without destroying the crystalline architecture. Ball-milling has been extensively researched in recent years due to its inexpensive cost, wide range of application, and small energy requirement for output.

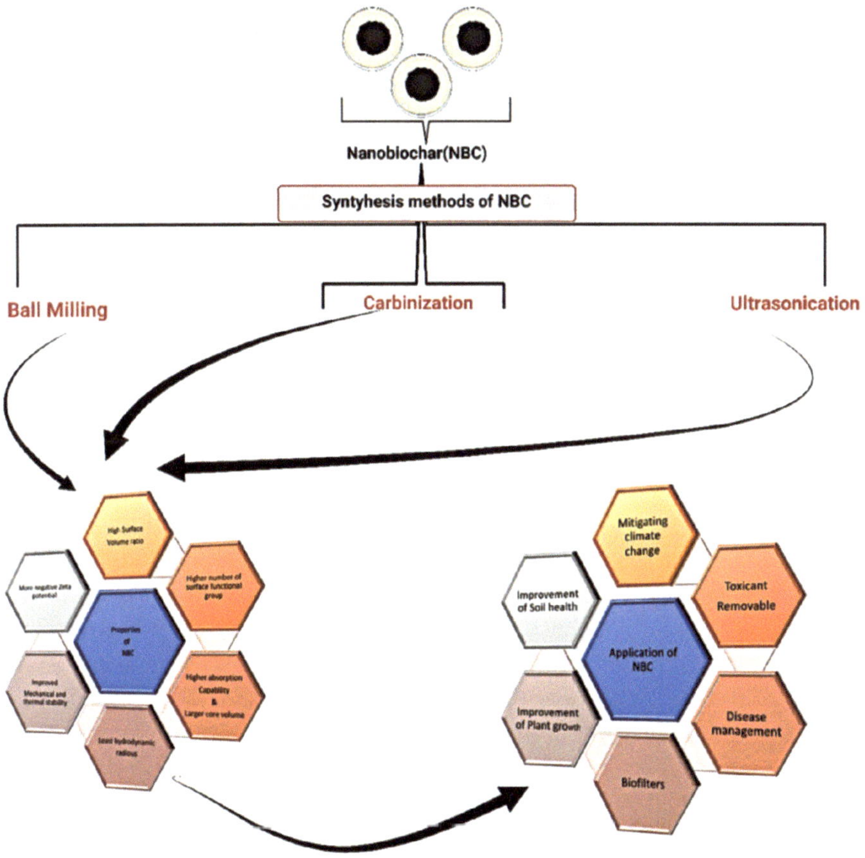

FIGURE 1.1 Overview, synthesis, types and application of nano biochar.

1.2.1 BALL MILLING

This technique is mostly applied by scientists in order to synthesize nano biochars. In the ball-milling procedure, it contains stainless balls and agate that have been employed. The material and ball size influence the mass of the balls used throughout milling. The weight of the ball depends upon the feature of the material and the ratio of stainless-steel balls and agate balls, like 30–45 g of stainless-steel balls and 180 g of agate balls were used in nano biochar synthesis from waste of sugarcane, waste of wood, and bamboo waste products. Several other factors like the type of milling, speed cycles of rotation, duration of milling are directly influencing the particle diameter of NBC. In a recent investigation, 100 g of agate balls were employed in the ball-milling process for 12 hours to produce magnetic wheat straw NBC. Thorough investigations that establish precise correlations among material, temperatures, and other characteristics with variations in particle diameter, however, are lacking (Shah et al., 2015).

1.2.2 ULTRASONIC METHOD OF NBC SYNTHESIS

The production of NBC can also be accomplished by the use of ultra-sonication. During the sonication process, a probe tip is fixed below the surface of the solution (Oleszczuk et al., 2016). This is the location where biochar granules are fragmented, and where shockwaves exfoliate the graphite-like structure of biochar (Li et al., 2013). The primary method includes the usage of ultrasound-induced microjets and shockwaves to generate new pores, unblock existing pores, and alter the structure of the biochar, which eventually results in an increase in microporosity (Sajjadi et al., 2020). The research conducted by Liu et al. (2018) showed that ultrasonic waves can dissociate fine fragments that are clinging to the surface or entrenched in the pores of the parent biochar, and that longer sonication times may result in the production of more NBC. According to the findings of Dong et al. (2018), ordinary biochar that was subjected to ultrasonic treatment was able to effectively synthesize NBC with a particle size of 426 nm. In the study by Song et al. (2019), ultrasonic work was performed for thirty minutes on bulk biochars derived from various source materials. As a result, NBC was obtained with a particle size of less than one hundred nanometers.

1.2.3 *OTHER SYNTHESIS METHOD OF NANO-BIOCHAR*

Despite the benefits of ball-milling in creating nanoparticles, researchers have implemented traditional ways of grinding materials before screening to result in aqueous suspension to create NBC. For instance, Li et al. (2017) created pitch pine NBC by smashing macro biochar at high speed and sieving the resulting material using a stainless-steel mesh (pore size = 74 nm). The biochar colloids were then dissolved in water, and the collected NBC was taken out of the supernatant. In order to disintegrate biochar on a smaller scale, researchers have also employed a disc mill rather than a ball mill. In one experiment, the pulverized biochar was dissolved in ethanol, centrifuged, dried, and the nano-fraction was gathered. Liu et al. (2018) prepared goethite-modified peanut shell biochar using a similar technique, sifting out the biochar powder using a mesh with pores that ranged in size from 75 to 150 nm. To settle the particles larger than 100 nm, the colloidal biochar was dissolved in water for a period of 24 hours. The nano biochar-containing suspension from the residual suspension was pipetted out after it had been centrifuged. The main difference between the manufacture of grass, wheat straw, and rice husk nano biochar was the mesh size, which was 500 nm and 50 nm, respectively (Oleszczuk et al., 2016).

Softwood was microwave pyrolyzed for 25–50 minutes at three distinct microwave power levels, 2,100, 2,400, and 2,700 W, in a study by Wallace et al. (2019). Due to the direct production of NBC without any additional processing, this method may be more practical. One kilogram of biomass was mixed with a microwave absorber (10 wt%) and pyrolyzed for 60 minutes in a nitrogen atmosphere using the microwave absorber to raise the reaction temperature.

1.3 PROPERTIES OF NBC

1.3.1 *SURFACE AREA*

Biochar is very surface-area dense, which is affected by the quantity of volatile organic compounds (VOCs) emitted during the carbonization process. Research has shown that certain forms of biochar, when subjected to appropriate ambient temperatures, exhibit quite significant surface areas. However, it has been demonstrated that the surface area changes significantly with respect to the temperature of the pyrolysis process. NBC synthesis occurred at 300 and 400°C, whereas NBC developed at 450 and 600°C, representing much higher surface areas. It is expected that the deaeration of biomass material will result in the formation of pores on the material's surface, which will lead to an elevated surface area of the substance. However, the amount of hemicellulose present in the material determines the surface area, and remodeling strategies are being implemented to improve the surface areas of biochars that are rich in lignin (Ramanayaka et al., 2022).

1.3.2 OUTPUT, VOLATILE MATTER, AND ORGANIC CARBON

When the temperature exceeds about 950°C, combustible stuff is liberated from the feedstock. A softwood NBC approximate content (in weight percent) was 25% high volatility matter (VM) and 71.3% organic carbon (Ramanayaka et al., 2022). The amount of VM released depends on the kind of feedstock, and some softwood feedstocks, like cinnamon, might emit greater VM than other softwood feedstocks. It is fascinating to learn that pinewood NBC, which has resin hardwood, has been shown to have 97% VM and only 1% fixed carbon (Naghdi et al., 2016). According to published research, macro-scale softwood biochar, such as dendro biochar, has a fixed carbon concentration of 63.8% and 19.7% of VM. The given findings show that, in relation to particle size, there has been no appreciable change in the amounts of FC and VM. The elemental composition of macro and NBC, moreover, can change significantly based on particle size (Oleszczuk et al., 2016). Moreover, a strong association between the color and the output of biochars at the nano size and microscopic scale was observed (Song et al., 2019). The same pattern has been seen in nano and micro fractions. More outputs were obtained from darker-colored suspensions than from lighter-colored suspensions. For instance, hardwood like pinewood biochar solutions (bright) produced 1.43%, while softwood like barley grass biochar solutions (dark) produced 20.5% (Song et al., 2019).

1.4 WATER HOLDING CAPACITY

The hydrological characteristics of biochar are clearly affected by the pyrolysis procedure and the porosity of feedstocks. The functional groups found on the surface of feedstocks decrease as pyrolysis temperature increases, which has the effect of reducing water binding capability and increasing hydrophobic nature (Ramanayaka et al., 2022). The propensity for water increases with the number of functional groups. As a result, there is no discernible variation in water-holding capacity as the size of particles decreases dependent on the pyrolysis temperature. The permeability and arrangement of the pores also play a role in a substance's capacity to hold water.

Pinewood NBC reported exhibiting water retention capabilities of 9.75 and 2.7%, respectively, when produced by pyrolysis at 525 and 400°C (Yu et al., 2013). However, because there were so few investigations, the comparison of the scant information was insufficient to draw any firm conclusions. Because of weak interconnection and coagulation of the pores with residual materials, such as tar, depending on the feedstock, biochars with relatively low temperature also include pores that are simply too tiny to store a substantial quantity of water (Das & Sarmah, 2015).

1.5 TYPES OF THE NANO BIOCHAR

Figure 1.2 illustrates the various types of biochar-based nanostructures. According to the different kinds of nanostructures being used in NBC, they can be classified into three categories: fully functioning nanostructures encapsulated in biochar (such as graphene, graphene oxide, chitosan, carbon nanotubes, ZnS nanocrystals, graphitic C_3N_4, etc.); oxide/hydroxide biochar nanocomposites; and magnetic biochar nanocomposites.

Several synthetic procedures are used to create these composite materials. Every formulation technique includes two processing steps for the feedstock, called pre- and post-treatment.

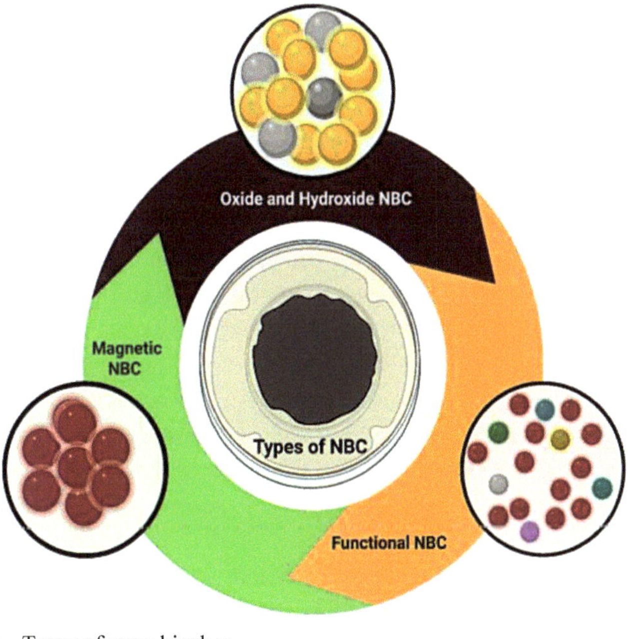

FIGURE 1.2 Types of nano biochar.

1.5.1 OXIDE/HYDROXIDE MEDIATED NANO-BIOCHAR

Oxide and hydroxide NBC are basically prepared through three procedures: fortification of target elements by the process of bioaccumulation; pre-treatment of feedstock via metal salt; and post-pyrolysis of nanomaterial oxide. Among the three procedures, the first and second procedures are dependent on the pre- and post-pyrolysis of feedstock along with metal fortification, while the third procedure is totally dependent on the addition of oxide and hydroxide nanomaterials into pyrolyzed biochar derived from feedstock (Ramanayaka et al., 2022).

1.5.2 FUNCTIONAL NANO BIOCHAR

Reasonable nanomaterial composites can be produced by coating biochar with functional nanomaterials such as graphene and its oxides, chitin, carbon nanotubes, stacked double hydroxides, and so on. These nanostructured materials have the capacity to thoroughly eliminate a wide variety of pollutants from different sources (Ramanayaka et al., 2022).

1.5.3 MAGNETIC NANO BIOCHAR

Two methods of pre-treating biomass with iron ions or chemically coprecipitating iron oxides onto biochar can be used to create magnetic nanoparticles on biochar, while considering the challenge of separating from the aqueous solution. Several methods might encapsulate nanosized magnetic iron oxide, such as $CoFe_2O_4$ particles (He et al., 2020), Fe_3O_4 particles (Peterson et al., 2021), and α-Fe_2O_3 particles (Wang et al., 2020), onto the interface of biochar, providing the initiation of magnetism and the accessibility of catalytic activity for iron oxide for the elimination of impurities. Biochar's magnetic properties make it easier to separate from solutions and improve its capability for adsorption (Ramanayaka et al., 2022).

The removal of pollutants has been accomplished with good efficacy by composites with increased features including effective surface area, active site accessibility, catalytic breakdown capability, pore-size distribution, pore volume, and ease of separation. Several experiments revealed that such composite materials were more effective at adsorbing a variety of contaminants from aqueous medium (Zhau et al., 2020). Also, these nanocomposites that contained a variety of reductive, oxidative, and catalytic nanoparticles (such as graphitic C_3N_4 and nanoscale zero-valent iron, etc.) concurrently bind and disintegrate several hazardous compounds (Ramanayaka et al., 2022).

1.6 APPLICATION OF NBC

NBC has several uses such as biofilter, energy production, enhancement of soil health, disease management, phytoremediation, remediation of toxic substances from soil as well as water, and improvement of plant growth, among others (Figure 1.3). NBC is an ecologically sustainable and cost-effective bioremediation technology. Other processes are costly and pollute the environment. Its specific features make it suitable for the formation of chemical immunosensors for toxin screening in water sources. NBC promotes soil microorganisms and removes pollutants. Despite its environmental benefits, NBC may have harmful effects. NBC may cause biochemical or physicochemical harm. We all know that nanomaterials such as regulated medicine delivery, agricultural applications, and nanochips have improved. Biochar-derived nanosensors may easily identify and analyze contaminants. NBC electrodes generate power. NBC is exclusively utilized in pot trials despite its many uses. MNBC needs further research in esterification, nano-sensor, and high-value applications to minimize pollution and make people's lives easier. As a result, nanomaterials constructed on biochar offer a great deal of potential for treating a variety of wastes and for use in a wide range of ecological purposes (Ramanayaka et al., 2022).

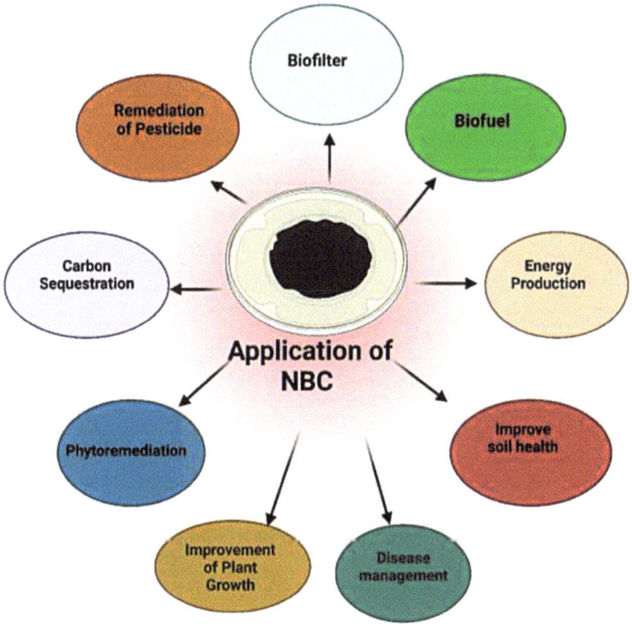

FIGURE 1.3 Application of nano biochar.

1.7 CONCLUSION

The chapter has mainly shown how nano and nano biochar can be synthesized, its type and its applicability to solve various problems of agriculture and the environment. Heavy metals, dyes, pharmaceuticals, and other organic and inorganic contaminants can all be effectively absorbed by NBC. Moreover, NBC has a variety of uses in nutrient recycling, pesticide remediation, fertilizer usage, soil improvement, eradicating the pathogenic effect, and more, a phenomenon commonly referred to as "The Biochar Effect." As recently proven, the interaction of the soil-plant-biochar system results in the biochar effect, which ultimately leads to biochar-induced alterations in soil microorganisms and biodiversity. Much more research is required; however, it has also been used in the fields of healthcare, electrode objects, supercapacitors, and biosensors. Despite their many benefits, the biological as well as ecological toxicity of NBC and nanocarbon materials must be dealt with because there is just a limited amount of research on their impact on numerous species of plants, mammalian cell systems, and soil microbial communities. Furthermore, it is important to evaluate the impact of the biochar base materials, particle size, variety of the PAH profiles, concentration fluctuation with pyrolysis temperature, and the quantity of pollutants. In addition, more study is required to comprehend the underlying mechanisms of the nano-similar efficacy of biochar on plant development, the microbial population, and the environment. Hence, research efforts in the future will be concentrated on investigating ways to improve the yield of biochar, the manufacturing of electrode supercapacitors, the reduction of clumping, and applicability in the medical field. Another area to research the possibilities of biochar is plant nanobionics, since these materials can be used as nanocarriers for the delivery of insecticides, nucleic acids, and fertilizers to crops and plants.

ACKNOWLEDGMENT

Financial support from the Strategic Academic Leadership Program of the Southern Federal University ("Priority 2030") is acknowledged by the authors.

KEYWORDS

- **carbon nanotubes**
- **chitosan**
- **graphene**
- **graphene oxide**
- **graphitic C3N4**
- **nano biochar**
- **nanotechnology**
- **reactive organic species**
- **volatile organic compounds**
- **volatility matter**
- **ZnS nanocrystals**

REFERENCES

Azadi, N., & Raiesi, F., (2021). Biochar alleviates metal toxicity and improves microbial community functions in a soil co-contaminated with cadmium and lead. *Biochar, 3,* 485–498.

Chen, H., Gao, Y., Li, J., Sun, C., Sarkar, B., Bhatnagar, A., Bolan, N., et al., (2022). Insights into simultaneous adsorption and oxidation of antimonite [Sb(III)] by crawfish shell-derived biochar: Spectroscopic investigation and theoretical calculations. *Biochar, 4*, 37.

Das, O., & Sarmah, A. K., (2015). The love–hate relationship of pyrolysis biochar and water: A perspective. *Sci. Total Environ., 512*, 682–685.

Dong, X., He, L., Liu, Y., & Piao, Y., (2018). Preparation of highly conductive biochar nanoparticles for rapid and sensitive detection of 17β-estradiol in water. *Electrochim Acta, 292*, 55–62.

Gong, Y., Wang, Y., Lin, N., Wang, R., Wang, M., & Zhang, X., (2022). Iron-based materials for simultaneous removal of heavy metal(loid)s and emerging organic contaminants from the aquatic environment: Recent advances and perspectives. *Environ. Pollut., 299*, 118871.

He, L., Yang, Y., Kim, J., Yao, L., Dong, X., Li, T., & Piao, Y., (2020). Multi-layered enzyme coating on highly conductive magnetic biochar nanoparticles for bisphenol A sensing in water. *Chem. Eng. J., 384*, 123276.

Hou, D., O'Connor, D., Igalavithana, A. D., Alessi, D. S., Luo, J., Tsang, D. C. W., Sparks, D. L., et al., (2020). Metal contamination and bioremediation of agricultural soils for food safety and sustainability. *Nat. Rev. Earth Environ., 1*, 366–381.

Huang, X., Zhu, S., Zhang, H., Huang, Y., Wang, X., Wang, Y., & Chen, D., (2021). Biochar nanoparticles induced distinct biological effects on freshwater algae via oxidative stress, membrane damage, and nutrient depletion. *ACS Sustain. Chem. Eng., 9*, 10761–10770.

Khan, A. H., Aziz, H. A., Khan, N. A., Hasan, M. A., Ahmed, S., Farooqi, I. H., Dhingra, A., et al., (2021). Impact, disease outbreak and the eco-hazards associated with pharmaceutical residues: A critical review. *Int. J. Environ. Sci. Te., 19*, 677–688.

Li, D. P., Ma, J., Xu, H. C., Xu, X. Y., Qiu, H., Cao, X. D., & Zhao, L., (2022). Recycling waste nickel-laden biochar to pseudo-capacitive material by hydrothermal treatment: Roles of nickel-carbon interaction. *Carbon Res., 1*, 16.
Li, H., Dong, X., da Silva, E. B., de Oliveira, L. M., Chen, Y., & Ma, L. Q., (2017). Mechanisms of metal sorption by biochars: Biochar characteristics and modifications. *Chemosphere, 178*, 466–478.
Li, R., Zhang, Y., Deng, H., Zhang, Z., Wang, J. J., Shaheen, S. M., Xiao, R., et al., (2020). Removing tetracycline and Hg(II) with ball-milled magnetic nanobiochar and its potential on polluted irrigation water reclamation. *J. Hazard Mater., 384*, 121095.
Li, W., Zhu, X., He, Y., Xing, B., Xu, J., & Brookes, P. C., (2013). Enhancement of water solubility and mobility of phenanthrene by natural soil nanoparticles. *Environ. Pollut., 176*, 228–233.
Lian, F., Yu, W., Zhou, Q., Gu, S., Wang, Z., & Xing, B., (2020). Size matters: Nano-biochar triggers decomposition and transformation inhibition of antibiotic resistance genes in aqueous environments. *Environ. Sci. Technol., 54*, 8821–8829.
Liu, G., Zheng, H., Jiang, Z., Zhao, J., Wang, Z., Pan, B., & Xing, B., (2018). Formation and physicochemical characteristics of nano biochar: Insight into chemical and colloidal stability. *Environ. Sci. Technol., 52*, 10369–10379.
Liu, L., Li, W., Song, W., & Guo, M., (2018). Remediation techniques for heavy metal-contaminated soils: Principles and applicability. *Sci. Total Environ., 633*, 206–219.
Liu, X., Liu, L., Leng, P., & Hu, Z., (2022). Syringa reticulata subsp. amurensis growth influencing the microbial diversity and heavy metal contents in mixed substrate of municipal sludge and construction waste. *Env. Pollut. Bioavail., 34*, 127–135.
Liu, Y., Luo, X., Zeng, Y., Wang, Q., Tu, W., Yang, C., & Mai, B., (2020b). Trophic magnification of short- and medium-chain chlorinated paraffins in terrestrial food webs and their bioamplification in insects and amphibians during metamorphosis. *Environ. Sci. Technol., 54*, 11282–11291.
Luo, Y., Wang, Y., Zhu, Y., Xue, M., Zheng, A., Han, Y., Yin, Z., et al., (2022). Ball-milled bismuth oxychloride/biochar nanocomposites with rich oxygen vacancies for reactive red-120 adsorption in aqueous solution. *Biochar, 4*, 21.
Lyu, H., Gao, B., He, F., Zimmerman, A. R., Ding, C., Huang, H., & Tang, J., (2018). Effects of ball milling on the physicochemical and sorptive properties of biochar: Experimental observations and governing mechanisms. *Environ. Pollut., 233*, 54–63.
Lyu, H., Gao, B., He, F., Zimmerman, A. R., Ding, C., Tang, J., & Crittenden, J. C., (2018). Experimental and modeling investigations of ball-milled biochar for the removal of aqueous methylene blue. *Chem. Eng. J., 335*, 110–119.
Lyu, H., Xia, S., Tang, J., Zhang, Y., Gao, B., & Shen, B., (2020). Thiol-modified biochar synthesized by a facile ball-milling method for enhanced sorption of inorganic Hg^{2+} and organic CH_3Hg^+. *J. Hazard Mater., 384*, 121357.
Naghdi, M., Taheran, M., Pulicharla, R., Rouissi, T., Brar, S. K., Verma, M., & Surampalli, R. Y., (2016). Pine-wood derived nanobiochar for removal of carbamazepine from aqueous media: Adsorption behavior and influential parameters. *Arab. J. Chem.* doi: 10.1016/j.arabjc.2016.12.025.
Oleszczuk, P., Ćwikła-Bundyra, W., Bogusz, A., Skwarek, E., & Ok, Y. S., (2016). Characterization of nanoparticles of biochars from different biomass. *J. Anal. Appl. Pyrolysis, 121*, 165–172.
Peterson, S. C., Jackson, M. A., Kim, S., & Palmquist, D. E., (2021). Increasing biochar surface area: Optimization of ball-milling parameters. *Powder Technol., 228*, 115–120.

Pratap, T., Chaubey, A. K., Patel, M., Mlsna, T. E., Pittman, C. U. Jr., & Mohan, D., (2022). Nanobiochar for aqueous contaminant removal. In: Mohan, D., Pittman, C. U., & Mlsna, T. E., (eds.), *Sustainable Biochar for Water and Wastewater Treatment* (pp. 667–704). Elsevier, Amsterdam.

Qin, Y., Li, G., Gao, Y., Zhang, L., Ok, Y. S., & An, T., (2018). Persistent free radicals in carbon-based materials on transformation of refractory organic contaminants (ROCs) in water: A critical review. *Water Res., 137*, 130–143.

Ramanayaka, S., Vithanage, M., Alessi, D. S., Liu, W. J., Jayasundera, A. C. A., & Ok, Y. S., (2020). Nanobiochar: Production, properties, and multifunctional applications. *Environ. Sci. Nano, 7*, 3279–3302.

Ren, S. Y., Xu, X., Hu, K. S., Tian, W. J., Duan, X. G., Yi, J. B., & Wang, S. B., (2022). Structure-oriented conversions of plastics to carbon nanomaterials. *Carbon Res., 1*, 15.

Sajjadi, B., Chen, W. Y., Mattern, D. L., Hammer, N., & Dorris, A., (2020). Low-temperature acoustic-based activation of biochar for enhanced removal of heavy metals. *J. Water Process Eng., 34*, 101166.

Samadi, A., Xie, M., Li, J., Shon, H., Zheng, C., & Zhao, S., (2021). Polyaniline-based adsorbents for aqueous pollutants removal: A review. *Chem. Eng. J., 418*, 129425.

Shah, U. V., Wang, Z., Olusanmi, D., Narang, A. S., Hussain, M. A., Tobyn, M. J., & Heng, J. Y., (2015). Effect of milling temperatures on surface area, surface energy and cohesion of pharmaceutical powders. *Int. J. Pharm., 495*, 234–240.

Singh, M., Ahsan, M., Pandey, V., Singh, A., Mishra, D., Tiwari, N., Singh, P., Karak, T., & Khare, P., (2022). Comparative assessment for removal of anionic dye from water by different waste-derived biochar vis a vis reusability of generated sludge. *Biochar, 4*, 13.

Song, B., Chen, M., Zhao, L., Qiu, H., & Cao, X., (2019). Physicochemical property and colloidal stability of micron- and nano-particle biochar derived from a variety of feedstock sources. *Sci. Total Environ., 661*, 685–695.

Sun, Y., Lyu, H., Cheng, Z., Wang, Y., & Tang, J., (2022). Insight into the mechanisms of ball-milled biochar addition on soil tetracycline degradation enhancement: Physicochemical properties and microbial community structure. *Chemosphere, 291*, 132691.

Wallace, C. A., Afzal, M. T., & Saha, G. C., (2019). Effect of feedstock and microwave pyrolysis temperature on physio-chemical and nano-scale mechanical properties of biochar. *Bioresour. Bioprocess, 6*, 33.

Wang, K., Sun, Y., Tang, J., He, J., & Sun, H., (2020). Aqueous Cr (VI) removal by a novel ball-milled Fe0-biochar composite: Role of biochar electron transfer capacity under high pyrolysis temperature. *Chemosphere, 241*, 125044.

Wang, Y., Wang, L., Li, Z., Yang, D., Xu, J., & Liu, X., (2021). MgO-laden biochar enhances the immobilization of Cd/Pb in aqueous solution and contaminated soil. *Biochar, 3*, 175–188.

WHO, (2015). *Lead Poisoning and Health.* World Health Organization, Geneva Pure Earth; Green Cross Switzerland. World's worst pollution problems.

Xiao, Y., Lyu, H., Tang, J., Wang, K., & Sun, H., (2020). Effects of ball milling on the photochemistry of biochar: Enrofloxacin degradation and possible mechanisms. *Chem. Eng. J., 384*, 123311.

Yu, O. Y., Raichle, B., & Sink, S., (2013). Impact of biochar on the water holding capacity of loamy sand soil. *Int. J. Energy Environ. Eng., 4*, 44.

Zhang, J., Hu, X., Zhang, K., & Xue, Y., (2019a). Desorption of calcium-rich crayfish shell biochar for the removal of lead from aqueous solutions. *J. Colloid Interface Sci., 554*, 417–423.

Zhang, P., Xue, B., Jiao, L., Meng, X., Zhang, L., Li, B., & Sun, H., (2022). Preparation of ball milled phosphorus-loaded biochar and its highly effective remediation for Cd- and Pb-contaminated alkaline soil. *Sci. Total Environ., 813*, 152648.

Zhang, X., Wells, M., Niazi, N., Bolan, N., Shaheeng, S., Hou, D., Gao, B., et al., (2022). Nanobiochar-rhizosphere interactions: Implications for the remediation of heavy-metal contaminated soils. *Environ. Pollut., 299*, 118810.

Zhang, Y., Zhao, G., Xuan, Y., Gan, L., & Pan, M., (2021). Enhanced photocatalytic performance for phenol degradation using ZnO modified with nano-biochar derived from cellulose nanocrystals. *Cellulose, 28*, 991–1009.

Zhou, Y., Xiao, J., Hu, R., Wang, T., Shao, X., Chen, G., Chen, L., & Tian, X., (2020). Engineered phosphorous-functionalized biochar with enhanced porosity using phytic acid-assisted ball-milling for efficient and selective uptake of aquatic uranium. *J. Mol. Liq., 303*, 112659.

CHAPTER 2

Benefits of Nano-Enhanced Biochar

KHALIL MSEDDI,[1] VISHNU D. RAJPUT,[2] TATIANA MINKINA,[2] PRADEEP KUMAR,[3] and FAIÇAL BRINI[4]

[1]*Department of Biology, Faculty of Science of Sfax, University of Sfax, Sfax, Tunisia*

[2]*Academy of Biology and Biotechnology, Southern Federal University, Rostov-on-Don, Russia*

[3]*Department of Botany, MMV, Banaras Hindu University, Varanasi, Uttar Pradesh, India*

[4]*Laboratory of Biotechnology and Plant Improvement, Center of Biotechnology of Sfax, Sfax, Tunisia*

ABSTRACT

Recently, biochar (BC) has received attention as a useful additive for soil, and its production has increased substantially on a worldwide scale. The synthesis of biochar occurs as a byproduct of the thermochemical process of biomasses that takes place in the absence of or with restricted access to oxygen. Applications using biochar as a component for combating pollution and climate change are quite recent. From 2014, research on the environmentally responsible production of nano-biochar (also known as nano-BC) for use in agriculture and soil has been carried out as a result of the development of nanotechnology. Micro-sized BC with diameters of less than a micrometer (μm) and up to a nanometer (nm), also known as "dissolved" and "nano-BC," is created during the carbonization process. A recent study produced nano-BC with a diameter of less than 5 nm. In this review paper, we reported the benefits of Nano-enhanced Biochar in all use

Nanomaterials and Nano-Biochar in Reducing Soil Stress: An Integrated Approach to Sustainable Agriculture. Vishnu D. Rajput, & Tatiana Minkina (Eds.)

domains following the enhanced physical-chemical properties compared to bulk biochar.

2.1 INTRODUCTION

As a fertilizer for the soil, biochar has seen a significant rise in popularity and is now widely manufactured around the world. It is created through the thermochemical conversion of biomasses when oxygen is not present or is only partially present. There are many uses for biochar, a carbonaceous byproduct of lignocellulosic biomass pyrolysis. Important biochar applications include restoring contaminated water, removing pollutants, retaining nitrogen, boosting soil fertility, and promoting plant growth. Conversely, biochar modifications can improve its applications by altering its properties. Seeking to influence biochar, on the other hand, can bring about a shift in basic characteristics and, as a result, enhance its applicability. Nanobiochar is produced by decreasing biochar to the nanometer scale; it possesses high porosity, a huge surface area, interface active area, as well as enhanced surface functionalities. The term "nanobiochar" was coined to describe this product. The distinction between bulk-biochar and Nano-biochar can be seen in structural and physicochemical properties. Numerous elemental compositions, aromatic/polar natures, cation exchange capacity (CEC), pH, pore properties, specific surface area (SSA), and Zeta potential can all be present in Nano-biochar. The biomass pyrolysis conditions used, such as milling time and temperature, are primarily responsible for determining the size of biochar. Nanobiochar is widely used for contaminants detection, nutrient retention, and pollutant adsorption. Making of nanobiochar in the process of generating biochar from lignocellulosic materials, such as wood products, crop wastes, and animal waste products. Several other thermochemical procedures such as slow pyrolysis, gasification, rapid pyrolysis, torrefaction, and carbonization (hydrothermal or flash), are being used. These procedures are combined in order to develop biochar. Slow pyrolysis, which takes place in an atmosphere devoid of oxygen, is one of the most common approaches for fabricating biochar. This process typically results in a 35 percent dry biomass yield and produces a minimal quantity of bio-oil.

One of the many environmental applications of nanobiochar that is now the focus of study is phytoremediation. This broad term encompasses an extensive series of practices, including waste management, waste recycling, sewage treatment, and many more. Carbon sequestration, energy production, and phytoremediation are some of the many possibilities of this technology. Biochar has a high potential to remove toxins from water. This is due to the fact that it can be made from a diverse selection of feedstocks, that it is inexpensive, and that its chemical and physical surface features are beneficial. In addition to enhancing the physical and chemical qualities of biochar, the production of biochar-based nanocomposites makes it possible to produce a unique composite and to combine the therapeutic benefits of biochar with those of other nanomaterials. This integration of biochar's beneficial properties with those of other nanomaterials is called synergy. In addition, biochar can reduce pollution, which can lead to major health and environmental issues, by sequestering hazardous compounds like pesticides and immobilizing metals. This process can be accomplished by the utilization of biochar. As a result of this, nanobiochar has been seen as a potentially useful instrument for the bioremediation of a wide variety of pollutants.

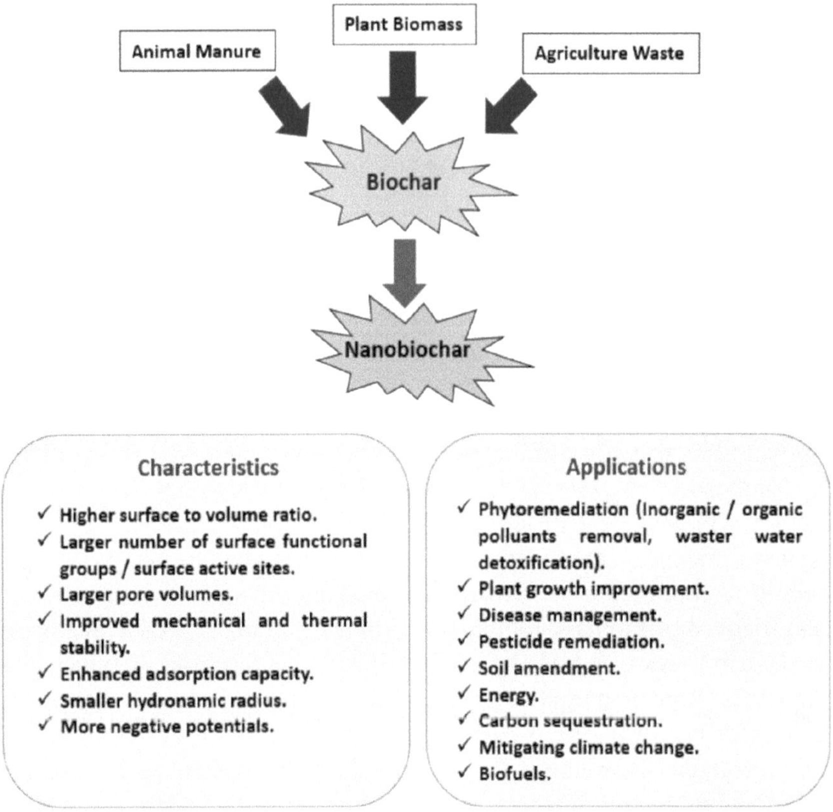

FIGURE 2.1 Characteristics, features, and applications of biochar and nanobiochar.

The organic substance known as Nanobiochar has been approved for use in agriculture for a number of purposes, including microbial growth support, disease prevention, pesticide cleanup, fertilization, and plant growth enhancement. Furthermore, it has been demonstrated that Nanobiochar boosts plant biomass, soil effectiveness, and productivity. It has also been demonstrated that the use of Nanobiochar in agricultural soil decreases the necessity of chemical-based fertilizers, particularly nitrogen fertilizers, reducing nitrogen leakage into groundwater. Because of its high charge density, nanobiochar not only enhances soil fertility, permeability, and water retention, but it also transports significant amounts of nutrients, boosting crop yield. In soil tainted with heavy metals, the use of nanobiochar significantly boosts microbial diversity, including bacteroidetes and actinobacteria (Figure 2.1). The objective of this work is to collect recent research releasing the benefits of nanobiochar in various fields of use.

2.2 BULK BIOCHAR VS. NANOBIOCHAR

It is necessary to delve deeper into the manufacturing of nano-biochar, as well as its classification and applicability, in order to provide a definition and differentiate it from various other particles formed from biomass. The terms biochar, nano-biochar, bulk biochar, fly ash, and soot are the most likely to cause confusion regarding the definition of nano-biochar presented in this review. Both nano-biochar and bulk biochar can be derived from biochar, which is created by the carbonization of biomass in an oxygen-deprived environment. All traditional biorefineries, including torrefaction, pyrolysis, and gasification, operate in an oxygen-deficient environment. Particle size variation is considered to be the main distinction between bulk and nano biochar. Fly ash and soot, on the other hand, are by-products of the combustion of different feedstocks, including biomass. It is suggested that a key indicator for differentiating biochar from fly ash and soot is how oxygen-deficient versus well-oxygenated their initial formation processes were. It is important to note that while biochar and soot are carbon-rich, fly ash produced by biomass combustion primarily comprises mineral compositions (such as SiO_2 and Al_2O_3) and alkali and alkaline earth metallic species (AAEMs) (Rajamma et al., 2009). Due to variations in feedstocks, this can be altered. For example, it has been reported that fly ashes can have 40% carbon content (Rajamma et al., 2009), which is comparable to biochar made from switchgrass pyrolysis and corn stover gasification (Brewer et al., 2009).

It is advisable to continue researching these compositional variations and/or similarities among biomass obtained from fly ash, soot, and biochar, as well as how they affect its characteristics and utility.

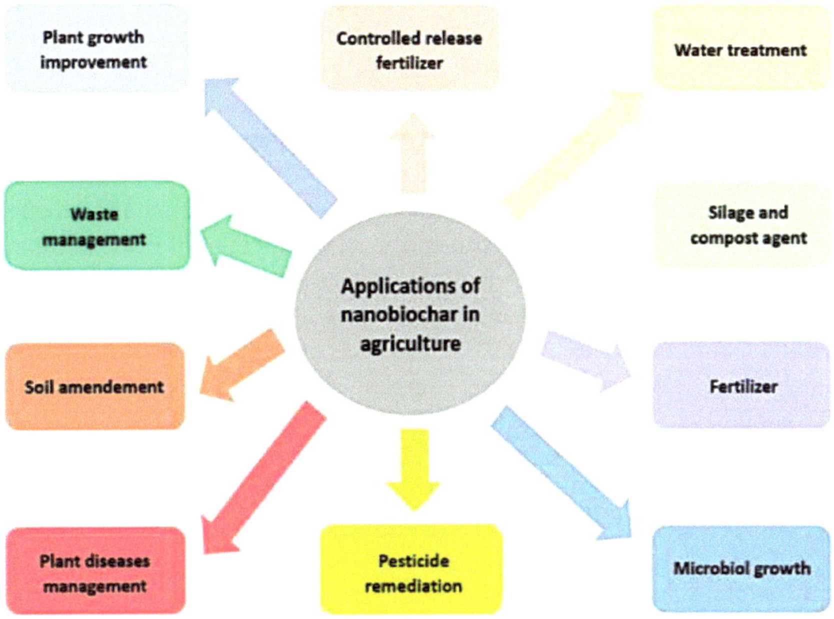

FIGURE 2.2 Utility of nano biochar in crop management.

2.3 NANOBIOCHAR AND BIOENERGY INNOVATION

Nanobiochar is defined as biochar with a particle size at or below 100 nm, despite the fact that the literature has categorized a wide range of particle diameters as nanobiochar, as illustrated in Figure 2.2. According to the ranges of particle diameter indicated in recent research, nano-biochar has been defined as biochar with a particle diameter within 0 and 600 nm. In fact, the term "nano-biochar" is used to describe the typical diameters of biochar, which range from 4 to 5 to about 500 nm. Due to the irregular range of particle diameters, it has been difficult to assess the effectiveness of distinct nano-biochars. This issue must be addressed in future research because the specific surface area of particles is strongly influenced by their size and shape. It is crucial to rapidly transition to a scenario where the planet's climate is carbon neutral or carbon negative in order to reduce the impact that human activities have on the natural atmosphere (Kirchherr et al.,

2017). Utilizing the systematic approach of circular economic development is a popular strategy for mitigating the negative impacts of human activities on the natural environment. This approach promotes economic growth while significantly reducing environmental impacts. Promoting the concept of environmental responsibility, the use of renewable energy sources such as wind, solar, biofuels, and geothermal energy is highly recommended. It is widely believed that the production of biofuels and biochemicals is an environmentally friendly technique that not only provides alternatives to the burning of fossil fuels but also reduces pollution caused by improper disposal of biowaste (Ubando et al., 2020).

The greatest obstacle to the efficient adoption of biofuel and bioenergy production is economic competitiveness. Realizing the full value of biomass resources and biorefining derivatives in a regenerative manner will boost the economic viability of producing biofuels and bioenergy while reducing pollutant emissions (Song et al., 2021). One of the most advanced biofuels, biochar, has shown that it may supplement solid fuel in the harvesting of bioenergy while solving critical environmental challenges. It can be produced using a variety of procedures including torrefaction, pyrolysis, and gasification at various temperatures and times (Song et al., 2021; Meyer et al., 2011). Together with variable levels of liquid and gaseous streams (such as bio-oil and syngas), biochar can be produced as the primary output or as a by-product of a process (Fang et al., 2020). Assuming a biochar output of 10%, the total production from biomass fast pyrolysis is approximated to be 220 million tons, and it is anticipated that as the desire for goods derived from biorefineries increases over the course of the next 30 years, this quantity will increase at a rapid rate (Woolf et al., 2010). Because of the intrinsic features that it possesses as a form of alternative energy, biochar is an excellent material or supplement for carbon sequestration, soil amendments, recycling of waste materials, and sewage remediation (Woolf et al., 2010; Lin et al., 2018).

The synthesis of catalysts and electrodes is a much more effective method to improve the utilization of biochar (Wang et al., 2019). The usage of biochar is increasing, which can speed up the transition to a carbon-neutral economy and boost the efficiency of biomass valorization (Glaser et al., 2009). Biochar is viewed as an intriguing source to produce carbon-based materials such as carbon nanotubes and graphene due to its advantages of having a low production cost, being widely accessible, possessing customizable features, and being sustainable (Qian et al., 2015).

2.4 APPLICATIONS OF NANOBIOCHAR IN MATERIAL SCIENCE, ENERGY, AND CATALYSIS

Additionally, nanobiochar has been applied as a sustainable substance in a number of industries, including energy, fuel, catalysis, biomedicine, and material science. Bio-nano additives are preferred over nano metal additives due to their non-toxic nature and play a significant role in reducing harmful emissions from vehicles when used with fuel. In their study, Mirbagheri et al. (2020) discovered that adding nanobiochar particles to a blend of diesel, biodiesel, and ethanol had a positive impact on engine performance and exhaust emissions. Another study found that sugarcane nano-biochar (SNB), which is made from bagasse, produced nanoparticles that outperformed metal-based fuel additives (Safieddin et al., 2020). In addition, the effectiveness of nanobiochar and biochar nanocomposites as enzyme supports and acid catalysts has been thoroughly investigated. The factors governing efficiency were improved in nanobiochar-based biocatalytic systems by increasing loading capacity, surface area, and decreasing resistance to mass transfer. Due to their porous structure, higher surface-to-volume ratio, and increased number of surface functional groups, these have been demonstrated to be effective immobilization supporting materials in enzyme systems (Naghdi et al., 2018). Functional groups, pore characteristics, and active sites all generally improved significantly in biochar-based nanocomposites, which enhanced catalytic devastation ability and accelerated separation (Ramadan et al., 2020; Tan et al., 2016). Additionally, Jenie et al. (2017) proposed using oil palm fruit bunches as biomass to create magnetic nanobiochar, which could then be utilized as heterogeneous acid catalysts. Additionally, when used as a heterogeneous acid catalyst in an esterification reaction, sulfonated magnetic nanobiochar was discovered to be superior to commercial solid catalyst or homogeneous acid catalyst (Jenie et al., 2020).

Nanobiochar has also been utilized as a sensing substance in a number of electrochemical devices. The biochar-based paper immunosensor or immunosensing device was designed for the identification of microcystin-LR (MCLR) toxic substance in freshwater. This was accomplished by encapsulating highly conductive and dispersible nanobiochar particles with anti-MCLR antibodies on filter paper using the dipping-drying process (Yao et al., 2021). Furthermore, Jiang et al. (2020) also revealed that waste lignin was recently transformed into nanobiochar. This material has the potential to be employed as a substitute for carbon black in the process of fortifying styrene-butadiene rubber in the rubber sector. As a result, nanobiochar and biochar nanostructures have a broad range of applications; however, there is an immediate necessity for additional studies to gather more specific evidence. In contrast to the applications described above, Goswami et al. (2020) used biochar-based biomaterial in an up-flow packed bed bioreactor (UFPBBR) to cleanse wastewater from biomass pyrolysis. The treatment resulted in a 96.2% reduction in the amount of contaminants in the sewage. As a result, there are a huge number of potential uses for nanobiochar that should be investigated.

2.5 NANO-BIOCHAR FOR ORGANIC POLLUTANTS DEGRADATION

Nanobiochar can be utilized as a transporter of enzymes or biocatalysts to facilitate improved breakdown of compostable molecules for a variety of tasks due to its excellent mobility and adjustable surface composition. This can be done for a number of different reasons (Itoh & Takagi, 2020). Immobilization of laccase, a potentially useful oxidation biocatalyst, upon nanobiochar produced from a variety of pyrochars was the subject of a number of experiments that were carried out by Brar et al. (2010). Comparable procedures are performed in order to manufacture laccase that is accompanied by nanobiochar. These procedures include an acidic pretreatment to introduce OH groups, oxidation to introduce -COOH groups, and enzyme immobilization as the last step (Naghdi et al., 2017).

It is possible to achieve a higher level of immobilization by using cross-linkers such as glutaraldehyde or carbodiimide hydrochloride (EDAC) (Naghdi et al., 2018, Zhang et al., 2020). Furthermore, a hydrogel approach was presented that encapsulates laccase between nano-biochar and chitosan. They claimed that attaching laccase to nano-biochar in this manner improved the material's stability under both storage and thermal conditions (4–70°C). Many different pollutants have been degraded using nano-biochar supported laccase. In a study by Naghdi et al. (2017), it was used to remove 3-ethylbenzothiazoline-6-sulphonic acid (ABTS) and achieved degradation rates of >80% with reusability for 3 cycles at a similar removal efficiency. When diclofenac (initial concentration 2.5 mg L^{-1}) was removed by Lonappan et al. (2018) using nano-biochar supported laccase, it took less time—just 2 hours—than laccase supported by carbon nanotubes (six hours) (Xu et al., 2015), and 40% of its activity was still present after five cycles.

Bacillus subtilis cultivation demonstrated the antibacterial activity of the nano-biochar supported laccase, indicating the viability of this composite for the treatment of bacteria-contaminated water (Naghdi et al., 2019). Recently, laccase supported by magnetic nano-biochar was created and used to remove bisphenol A (Zhang et al., 2020). When the initial concentration was 25 mg/L, the composite could remove all the bisphenol A (BPA) in 75 min, and 85% of the removal efficiency was still present after 7 cycles (Zhang et al., 2020). The most effective catalyst, according to experts, is nano-biochar supported laccase for the treatment of water contaminated with deoxyribonucleic acid (DNA). Lian et al. (2020) tested the adsorption and degradation properties of nano-biochar supported laccase to antibiotic resistance genes using DNA molecules and two ARGs from a natural river. Compared to 1.5 mg/g for bulk biochar supported laccase, the composite led to an adsorption of calf thymus DNA at 200 mg/L. The former composite showed that DNA molecules can fractionate, whereas bulk biochar supported laccase did not show any DNA molecule decomposition.

2.6 NANOBIOCHAR FOR ENVIRONMENT PURPOSES

Previous researchers have thoroughly investigated and demonstrated the use of macro-biochar in environmental applications. However, nanobiochar is now being researched for a number of environmental applications, including waste management, pollutant removal, wastewater treatment, and carbon sequestration. It has been established that carbon dioxide produced during the combustion of biomass, which includes plant matter or residues, is ultimately used by plants, making this a carbon-neutral source of energy. When biochar is applied to agricultural fields, the carbon is highly resistant to further deterioration and is sequestered for a very long time, on the order of centuries or millennia. As a result, biochar is thought of as a carbon-negative energy source that both produces energy and sequesters carbon. Biochar has become a promising technology in relation to various environmental problems, such as the problem of climate change (Timmons et al., 2017). Additionally, it is currently being researched for the generation of sustainable energy and electrode applications, in addition to its improved properties using various pre- and post-modification techniques (Ramanakaya et al., 2020).

Due to its greater adsorption capacity, nanobiochar is now known, among its many uses, as a good adsorbent. The effectiveness of nanobiochar and its sorption properties for removing pollutants like pharmaceuticals, steroid hormones, pesticides, toxic metals, and personal care products has been well established (Taheran et al., 2018). Through a variety of human activities, these micropollutants are easily dispersed in the atmosphere and pose a risk to both human health and the environment. Contaminants enter the environment through improper wastewater discharge, waste disposal, and the use of agrochemicals (Ramanakaya et al., 2019). Nanobiochar and biochar modified with nanominerals showed exceptional adsorption capability for the removal of a variety of contaminants when compared to regular biochar (Ma et al., 2019; Samsudin et al., 2019). Additionally, by adsorbing harmful chemicals like pesticides and immobilizing metals, biochar can eliminate pollution that could cause significant environmental and health problems (Cemansky, 2015). The application and simulation of biochar (BC) and nanomaterials (NMs) on decontaminating soil with potentially toxic elements (PTEs) have been reported as a perfect way to combat soil pollution (Rajput et al., 2022a). As a result, nanobiochar has been regarded as a potentially effective tool for the bioremediation of a variety of contaminants.

2.7 NANOBIOCHAR FOR PHYTOREMEDIATION OF HEAVY METALS

Because of its well-known adsorption properties, biochar has been used to remove pollutants through adsorption. Additionally, the ability of nanocarbons like black, activated, or engineered nanomaterial carbon to adsorb natural organic matter has been noted (Shen et al., 2019). Several investigators have discovered that nanobiochar is also very efficient in eliminating pollutants such as heavy metals, pharmaceuticals/drugs, dyes, and other contaminants with biochar (Ma et al., 2019; Samsudin et al., 2019). However, according to Naghdi et al. (2019), after three hours of contact time with nanobiochar derived from pine wood and having an average particle size of 60 nm, 95% of the medicine carbamazepine (CBZ) was eliminated from water using nanobiochar. In soil that had been polluted with cadmium, the implementation of nanobiochar was demonstrated to enhance plant biomass, seed germination, the stability of cadmium (Cd) ions, as well as the wide range of Actinobacteria and Bacteroidetes in the soil.

These results were reported in Liu et al. (2020)'s study. Additionally, soil Cd that was available was greatly reduced. In a study by Ramanayaka et al. (2019), nanobiochar was created using a top-down method and dendro biochar made from waste to remove a variety of contaminants from aqueous media, including antibiotics, agrochemicals, and potentially harmful elements. Results showed that glyphosate (GL), oxytetracycline (OTC), and Cd could all be removed using graphitic nanobiochar, which is derived from dendro biochar (II). According to another study by Li et al. (2018), ZnO/ZnS modified nanobiochar had the ability to adsorb Cu(II), Pb(II), and Cr(VI) with maximum absorption abilities of 135.8, 91.2, and 24.5 mg/g, respectively. These values were significantly higher than those for normal biochar (63.29, 27.05, and 15.23 mg/g, respectively). A number of other researchers have demonstrated that biochar nanocomposites have a significant ability to remove significant heavy metal contaminants, including arsenic, lead, copper, chromium, and cadmium, among others. These substances were also discovered to be successful at eliminating other contaminants, including harmful chemicals, dyes, medications, and antibiotics (Chausali et al., 2021).

Black tea waste (BTW and NBTW) and water hyacinth waste (BWH and NBWH) biochar and nanobiochar were used as potential low-cost and eco-friendly absorbents for the removal of heavy metals (HMs) from aqueous solutions (Elbehiry et al., 2022). Important micro- and macronutrients, as well as dangerous contaminants like potentially toxic metals and pesticides, may be effectively controlled by nanobiochar's mobilization and sorption (Rajput et al., 2022b).

2.8 NANOBIOCHAR FOR BIOREMEDIATION OF WASTEWATER

Heavy metal contamination in water is a serious environmental issue because of its accumulation and toxicity to living things in water bodies. In comparison to other organic and inorganic pollutants, heavy metals cannot be deteriorated, making their own reduction problematic (Ramadan et al., 2020). Due to its commercial applicability, adsorption from aqueous solutions/effluents is now the most widely used technique for the elimination of heavy metals (Guo et al., 2018). Other methods for the removal of heavy metals include ion exchange, chemical precipitation, and ion exchange. Adsorbents like activated carbon, flat iron oxide, silica gel, zeolite, and attapulgite, as well as carbon-based nanofibers, have a limited capacity for adsorption, are less effective economically, and have lower selectivity (Ramadan et al., 2020; Gan et al., 2015). Currently, biochar has shown promise as a contaminant adsorption agent, and it may also help reduce water pollution (Tan et al., 2015). Nanoparticle-infused biochar has improved properties and capabilities for a variety of uses, including carbon sequestration and wastewater treatment (Ramadan et al., 2020, p. 42). Particularly noteworthy for the removal of organic contaminants was the simultaneous adsorption and catalytic degradative action of catalytic material-coated biochar (Tan et al.,

2016). Conventional biochar, however, had a limited capacity to absorb ionic pollutants (Yao et al., 2013). Recent studies have shown various ways to create different types of nanobiochar to increase its ability to adsorb aqueous and ionic pollutants (Ramadan et al., 2020). The main action of nanobiochar on water pollution was the simultaneous adsorptive and degradative action of heavy metals like As, Cr, Pb, Cd, and Cu. According to Li et al. (2019), converting pure magnetic biochars (PMBCs) into ball-milled magnetic nanobiochar (BMBCs) made from wheat straw significantly improved the removal of tetracycline (TC) and Hg(II) from contaminated irrigation water. Additionally, TC and Hg(II) were 99% adsorbed by the action of BMBC within 12 hours. For TC and Hg(II), the removal capacities were 268.3 mg/g and 127.4 mg/g, respectively. Photosynthetic bacteria (PSB) loaded onto Fe_3O_4/biochar nanocomposites were used for wastewater bioremediation in another study by He et al. (2017).

The wastewater was cleaned with this PSB-composite mixture, which also removed 87.5% of NH_4^+ and 82.1% of chemical oxygen demand. It was determined that the PSB/Fe_3O_4/biochar composites were effective and could be used for wastewater bioremediation. Additionally, Li et al. (2017) used modified biochar materials such as $KMnO_4$ impregnated biochar (BC-MnOx), alkaline treated biochar (BC-NaOH), and magnetic biochar (BC-FeOx) treated with $FeCl_3$. Due to BC-maximum MnOx's sorption capacity of 81.10 mg/g and pseudo-second-order kinetics for Cd (II) absorption, the BC-MnOx composite's adsorption rate, which was the highest of all the composites, was discovered to be 14.46 g (mg h)$^{-1}$. Iron oxide permeated mesoporous rice-husk nanobiochars (IPMN), according to Nath et al. (2019), were the best for removing arsenic in the same order. Compared to the native milled rice husk (20%), it had an uptake capacity of >90% and could adsorb a significantly higher amount of As.

2.9 APPLICATIONS OF NANOBIOCHAR IN AGRICULTURE

An organic substance known as nanobiochar has been approved for use in agriculture for a number of purposes, including promoting microbial growth, managing diseases, removing pesticides from crops, and enhancing plant growth (Helal et al., 2019; Wang et al., 2017). In addition, nanobiochar was reported to enhance productivity (Alburquerque et al., 2013), soil efficiency (Liu et al., 2013), and plant biomass (Soja et al., 2013). Reduced use of chemical fertilizers, particularly nitrogen fertilizer, due to the use of nanobiochar in agricultural soil has also been found to reduce nitrogen leaching into groundwater (Orr et al., 2014).

2.9.1 *PLANT GROWTH ENHANCEMENT AND SOIL AMENDMENT*

The proportion of nutrients present in the soil has reduced dramatically as a result of a variety of anthropogenic activities like industrialization, deforestation, and excessive grazing. The decrease in crop production and plant development that resulted from nutrient deficiency has a severe adverse effect on the availability of foodstuffs (Ramadan et al., 2020). Chemical fertilizers have been overused in an attempt to maintain soil nutrients. In contrast, nutrient leaching and runoff reduced their bioavailability in soil and encouraged chemical-based fertilizers. This was the cause of soil pollution and negative effects on soil biota in many places (Bah et al., 2014). In addition to improving soil fertility, permeability, and water holding capacity, biochar also carried a significant amount of nutrients due to its high charge density, which increased crop yield (Liu et al., 2016; Helal et al., 2019). A vital nutrient for the growth of plants is phosphorus (P). Based on P adsorption, desorption, and transformation between the soil solids and solution, plants can obtain P as phosphate anions (H_2PO_4) from the soil solution (Helal et al., 2019; Debicka et al., 2016).

As a result, P adsorption and desorption affect soil P, which changes how P is utilized and taken up by plants. Biochar can be used as a soil-amending agent due to its soil-influencing qualities, as it has been shown to increase P and N availability (Zeng et al., 2013; Rawat et al., 2019) as well as decrease P fixation, nitrogen leaching, and N_2O emission (Dong et al., 2017; Pan et al., 2017). Salinity was found to be important for effective P absorption and to be improved as soil salinity rose. After treatment with 0.20% nanobiochar, the soil sample's Freundlich constant (KF) increased from its initial value of 64.60 mg/kg to 497.60 mg/kg (saline). Adsorption maxima of phosphorus (qmax) were found to rise in both non-saline and saline soils, having amounts of 9.43 mg/kg and 63.29 mg/kg, respectively (Mahmoud et al., 2020). As a result, biochar enhances plant growth and is crucial for managing soil nutrient levels for crop productivity. Once more, the addition of biochar to soil had an impact on soil characteristics like texture, porosity, particle size distribution, and density (Mangrich et al., 2015). As a result of its organic content, the soil's pH, cation exchange capacity (CEC), electrical conductivity (EC), and nutrient levels (C, N, P, etc.) have all increased (Liang et al.,

2006; Rondon et al., 2007; Amonette & Joseph, 2009) which has improved the efficiency of metal sorption (Beesley et al., 2010; Senthilkumar et al., 2020). Additionally, it significantly changed the microbial and nutritional status of the soil in the rhizosphere zone (Lehman et al., 2003), as well as the soil's air/oxygen content and water holding capacity (Mangrich et al., 2015; Laird et al., 2010).

Additionally, Chan et al. (2008) showed that using biochar decreased the likelihood of soil compaction by lowering the tensile strength of the soil. According to several studies, combining biochar with some cutting-edge technologies and engineered nanomaterials may produce better results. In a study by Saxena et al. (2014), treatment with biochar comprising water-soluble carbon nanoparticles dramatically boosted the development of wheat plants. According to Bopp et al. (2016), using alkaline biochar instead of acidic biochar significantly increased root growth and root biomass. Even though biochar and non-pyrogenic organic amendments (NPOA) support the process of plant development and growth, their special changing behaviors prevent them from being widely used for crops and other vegetation (Bonanomi et al., 2017). In addition, Yue et al. (2019) conducted studies to investigate the impacts of nanobiochar on rice seedlings with the goal of lowering metal toxicity (Cd^{2+}) and oxidative stress. They found that nanobiochar had a significant impact on lowering both of these factors.

2.9.2 NANOBIOCHAR A HIGH-QUALITY FERTILIZER

It is a high-performance organic fertilizer because biochar made from biomass contains higher concentrations of other macronutrients like phosphorus (P), nitrogen (N), sulfur (S), calcium (Ca), potassium (K), and magnesium (Mg) as well as micronutrients like manganese (Mn), iron (Fe), copper (Cu), and zinc (Zn) (Hossain et al., 2011). Biochar made from animal sources like broiler litter and sewage sludge was found to contain more total P and N than biochar made from plant biomass like wood and green waste (Bopp et al., 2016). When biochar was used in conjunction with other fertilizers, such as urea, production increased noticeably while N input was appropriately reduced (Joseph et al., 2013; Qian et al., 2014).

Additionally, the use of biochar composites containing minerals and organic materials can enhance the benefits of fertilizer use. Due to their ability to retain water and nutrients and their ease of use as nanofertilizers, biochar-derived nanocarbons have attracted a lot of attention (Manya, 2012;

Xiu et al., 2017). Furthermore, Zhang et al. (2020) reported that the development and productivity of wheat crops were noticeably improved by the application of biochar-carbon nanoparticles. Recently, Lateef et al. (2019) investigated how a biochar nanocomposite made from corn cobs helped protect the environment by turning waste into useful products and solved leaching issues with conventional fertilizers. However, more research is required to determine whether biochar-based fertilizer can effectively deliver nitrogen while immobilizing heavy metals (Ramadan et al., 2020).

2.9.3 MICROBIAL COMMUNITIES' ENHANCEMENT

Numerous agricultural uses of biochar have been investigated, including pesticide remediation, microbial growth support, and disease management (Saxena et al., 2016). The fact that the soil contains a variety of nutrients and ash is the fundamental aspect that provides the perfect habitation for microbes, which also in turn confers a variety of advantages upon the soil. It is the microbiological communities in plant soil that are responsible for maintaining soil quality, biogeochemical cycles, richness, biological control, as well as crop production. Such microbes are essential for the efficient functioning of the ecology found in plant soils (Thies et al., 2015). It has been discovered that biochar might alter the microbial biodiversity and have a significant impact on the microbial populations in the soil (Mangrich et al., 2015; Amonette & Joseph, 2009; Chan et al., 2007). Several features like pH, tensile strength, porosity, bulk density, surface area, CEC, and other fixed physical and chemical soil characteristics were also known to be modified by biochar (Mangrich et al., 2015; Amonette & Joseph, 2009; Lehmann et al., 2011). According to Liu et al. (2020), the use of nanobiochar led to a substantial rise in the microbiomes and organic matter of Actinobacteria and Bacteroidetes in soil containing Cd pollutants. On the other hand, the addition of nanobiochar led to a substantial decline in Proteobacteria diversity, which predominated in polluted areas. These observations were validated by previous studies conducted by Liao et al. (2019), which provided scientific validation regarding how biochar addition affected the population of soil microbes. These microbes helped in the absorption of carbon from plant-derived sources.

These reports revealed that the incorporation of nanobiochar led to a rise in the relative densities of members of the Bacteroidetes and Firmicutes phyla, such as *Sporomusa, Bacillus, Alicyclobacillus, Clostridium*, and

Desulfosporosinus, whereas resulting in a decrease in the loads of Proteobacteria, particularly Sphingomonas and Methylobacterium, both of which use. Additionally, Wu et al. (2020) mentioned the creation of a novel calcium-based magnetic biochar (Ca-MBC) that has dramatically transformed the species richness of microbes as well as raised the total number and population of bacteria, leading to a shift in structure. This change was brought about by the biochar's capacity to boost the number of bacteria and its demography. In addition, various research findings have demonstrated that the use of biochar can help decrease diseases in plants exacerbated by pathogenic microorganisms. This is because the use of biochar can stimulate systemic resistance among plants, which is prompted by relationships between all these soil microbes (such as Bacillus, Trichoderma, and Pseudomonas species) as well as plant roots. Consequently, the use of biochar can substantially reduce the incidence of plant diseases (Ramadan et al., 2020; Kolton et al., 2011).

2.9.4 NANOBIOCHAR FOR PESTICIDE SEQUESTRATION

Because of its superior porosity, which creates more adsorption sites than unaltered soil, biochar has also proven to be very effective in sorption and pesticide sequestration (Khalil et al., 2019). In this way, it may reduce pesticide desorption, mobility, leaching, and bioavailability in soil layers, leading to the remediation of pesticide-contaminated soils (Nag et al., 2011; Khorram et al., 2016). Biochar may facilitate a decrease in pesticide concentration in soil due to its high pesticide binding capacity (Jones et al., 2011). An additional study revealed that plants' uptake of the herbicide foreseen was reduced due to biochar's higher sorption capacity (Khorram et al., 2015). Because of their greater affinity for pesticides, these carbon nanomaterials can provide a long-term solution to pesticide-induced soil pollution and may be a candidate for pesticide remediation (Cernansky et al., 2015). More research and study will be needed in the future to fully understand biochar nanocomposites or nanobiochar and their ability to absorb various pesticides and herbicides.

2.10 CONCLUSION AND PERSPECTIVES

Most of the literature has shown how biochar nanocomposites or nanobiochar are used in environmental and agricultural problems. Nanobiochar has the ability to successfully absorb a wide variety of pollutants such as heavy metals, dyes, pharmaceuticals, and various organic and inorganic toxins. Furthermore, nanobiochar has a wide range of uses in recycling of nutrients, the remediation of pesticides, nutrient management administrations, soil amendment, and the eradication of the pathogenic effect, among other things. This phenomenon is widely referred to as "The Biochar Effect" (Elad et al., 2012). Chew et al. (2020) recently reported that the biochar effect is the outcome of the connection between the soil, plants, and biochar, which inevitably results in modifications to the activity and variety of microorganisms that are biochar-induced. Although a great deal further investigation is required, it has already been implemented in biosensing, electrode objects, supercapacitors, as well as medical services.

The biological and environmental toxic effects of nanobiochar and nanocarbon composites must be tackled because there have only been a few investigations done on their implications on a wide range of plant species, mammalian cellular systems, and soil microorganisms. This is even though both nanobiochar and nanocarbon materials have numerous advantages. In addition to this, it is necessary to evaluate the significance of the nanobiochar feedstock material, particle size, the variety of the PAH patterns, the fluctuation of concentration with temperature range, and the proportion of trace metals. Furthermore, there is a requirement for additional studies to be carried out in hopes of clarifying the processes that are behind the impacts that biochar's nanocomposite material has on the development of plant life, the microbial biomass, as well as the surroundings. Therefore, research efforts in the future will be concentrated on investigating ways to improve the yield of nanobiochar, the production of electrode supercapacitors, the reduction of agglomeration, and applications in the healthcare industry. Plant nanobionics is another area where biochar's potential is being investigated, as these materials can be used as nanocarriers for the release of pesticides, nucleic acids, and fertilizers to crops and plants.

ACKNOWLEDGMENT

The authors would like to acknowledge the financial support received from the Strategic Academic Leadership Program of the Southern Federal University ("Priority 2030").

KEYWORDS

- **agricultural application**
- **biochar**
- **climate change**
- **nanobiochar**
- **phytoremediation**

REFERENCES

Alburquerque, J. A., Salazar, P., Barron, V., et al., (2013). Enhanced wheat yield by biochar addition under different mineral fertilization levels. *Agronomy for Sustainable Development, 33*, 475–484.

Amin, F. R., Huang, Y., He, Y., Zhang, R., Liu, G., & Chen, C., (2016). Biochar applications and modern techniques for characterization. *Clean Technologies and Environmental Policy, 18*, 1457–1473.

Amonette, J. E., & Jospeh, S., (2009). Characteristics of biochar: Microchemical properties. In: Lehmann, J., & Joseph, S., (eds.), *Biochar for Environmental Management Science and Technology* (pp. 33–43). Earthscan, London.

Amrane, A., Mohan, D., Nguyen, T. A., Assadi, A. A., & Yasin, G., (2020). In: Amranne, et al., (eds.), *Nanomaterials for Soil Remediation: Part of Micro and Nano Technologies.* ISBN-10:‎0128228911, Elsevier.

Amusat, S. O., Kebede, T. G., Dube, S., & Nindi, M. M., (2021). Ball-milling synthesis of biochar and biochar–based nanocomposites and prospects for removal of emerging contaminants: A review. *Journal of Water Process Engineering, 41*, 101993.

Anupama., K. P., (2021). A comprehensive evaluation of inherent properties and applications of nano-biochar prepared from different methods and feedstocks. *Journal of Cleaner Production, 2021*, 128759.

Bah, A., Husni, M. H. A., Teh, C. B. S., et al., (2014). Reducing runoff loss of applied nutrients in oil palm cultivation using controlled-release fertilizers. *Advances in Agriculture*, 285387.

Bakshi, S., Banik, C., Rathke, S. J., & Laird, D. A., (2018). Arsenic sorption on zero-valent iron-biochar complexes. *Water Research, 137*, 153–163.

Beesley, L., Moreno-Jimenez, E., & Gomez-Eyles, J. L., (2010). Effects of biochar and greenwaste compost amendments on mobility, bioavailability and toxicity of inorganic and organic contaminants in a multi-element polluted soil. *Environmental Pollution, 15*, 2282–2287.

Bonanomi, G., Ippolito, F., Cesarano, G., et al., (2017). Biochar as plant growth promoter: Better off alone or mixed with organic amendments. *Frontiers in Plant Sciences, 8.*

Bopp, C., Christl, I., Schulin, R., & Evangelou, M. W., (2016). Biochar as possible long–term soil amendment for phytostabilization of TE–contaminated soils. *Environmental Sciences and Pollution Research, 23*, 17449–17458.

Brar, S. K., Verma, M., Tyagi, R. D., & Surampalli, R. Y., (2010). Engineered nanoparticles in wastewater and wastewater sludge–evidence and impacts. *Waste Management, 30*, 504–520.
Brewer, C. E., (2012). Biochar characterization and engineering. *Iowa State University.*
Brewer, C. E., Schmidt-Rohr, K., Satrio, J. A., & Brown, R. C., (2009). Characterization of biochar from fast pyrolysis and gasification systems. *Environmental Progress and Sustainable Energy, 28*, 386–396.
Buseck, P., Adachi, K., Gelencser, A., Tompa, E., & Posfai, M., (2012). Are black carbon and soot the same? *Atmospheric Chemistry and Physics, 12*, 24821–24846.
Cernansky, R., (2015). Agriculture: State–of–the–art soil. *Nature, 517*, 258–260.
Chan, K. Y., Van, Z. L., Meszaros, I., Downie, A., & Joseph, S., (2008). Using poultry litter biochars as soil amendments. *Soil Research, 46*, 437–444.
Chan, K. Y., Zwieten, L. V., Meszaros, I., Downie, A., & Joseph, S., (2007). Agronomic values of green waste biochar as a soil amendment. *Australian Journal of Soil Research, 45*, 629–634.
Chausali, N., Saxena, J., & Prasad, R., (2021). Nanobiochar and biochar-based nanocomposites: Advances and applications. *Journal of Agriculture and Food Research, 5*, 100191.
Chen, B., Chen, Z., & Lv, S., (2011). A novel magnetic biochar efficiently sorbs organic pollutants and phosphate. *Bioresources Technology, 102*, 716–723.
Chen, X., Zhou, B., Wang, Q., Tao, W., & Lin, H., (2020). Nano-biochar reduced soil erosion and nitrate loss in sloping fields on the Loess Plateau of China. *Catena, 187*, 104346.
Chew, J., Zhu, L., Nielsen, S., et al., (2020). Biochar-based fertilizer: Supercharging root membrane potential and biomass yield of rice. *Science of Total Environment, 713.*
Debicka, M., Kocowicz, A., Weber, J., & Jamroz, E., (2016). Organic matter effects on phosphorus sorption in sandy soils. *Archives of Agronomy and Soil Science, 62*, 840–855.
Dong, X., He, L., Liu, Y., & Piao, Y., (2018). Preparation of highly conductive biochar nanoparticles for rapid and sensitive detection of 17 β-estradiol in water. *Electrochimica Acta, 292*, 55–62.
Dong, X., Li, G., Lin, Q., & Zhao, X., (2017). Quantity and quality changes of biochar aged for 5 years in soil under field conditions. *Catena, 159*, 136–143.
Du, Y., Lin, W., & Glarborg, P., (2021). Particulate emissions from a modern wood stove–Influence of KCl. *Renew Energy, 170*, 1215–1227.
Elad, Y., Cytryn, E., Harel, Y. M., Lew, B., & Graber, E. R., (2012). The biochar effect: Plant resistance to biotic stresses. *Phytopathologia Mediterranea, 50*, 335–349.
Elbehiry, F., Darweesh, M., Al-Anany, F. S., Khalifa, A. M., Almashad, A. A., El-Ramady, H., & Elbasiouny, H., (2022). Using biochar and nanobiochar of water hyacinth and black tea waste in metals removal from aqueous solutions. *Sustainability, 14*(16), 10118.
Fang, Z., Smith, R. L., & Xu, L., (2020). Part I fundamentals of pyrolysis. *Production of Biofuels and Chemicals with Pyrolysis* (pp. 3–136). 10. Springer.
Gan, C., Liu, Y. G., Tan, X. F., et al., (2015). Effect of porous zinc–biochar nanocomposites on Cr (vi) adsorption from aqueous solution. *RSC Advances, 5*, 35107–35115.
Gao, L., Nie, L., Wang, T., Qin, Y., Guo, Z., Yang, D., & Yan, X., (2006). Carbon nanotube delivery of the GFP gene into mammalian cells. *ChemBioChem, 7*, 239–242.
Glaser, B., Parr, M., Braun, C., & Kopolo, G., (2009). Biochar is carbon negative. *Nature Geosci., 2*, 2. https://doi.org/10.1038/ngeo395.
Goswami, L., Pakshirajan, K., & Pugazhenthi, G., (2020). Biological treatment of biomass gasification wastewater using hydrocarbonoclastic bacterium *Rhodococcus opacus* in an

up-flow packed bed bioreactor with a novel waste-derived nano-biochar based bio-support material. *Journal of Cleaner Production, 256*, 120253.

Goswami, R., & Kumar, M., (2018). Removal of fluoride from aqueous solution using nanoscale rice husk biochar. *Groundwater Sustainable Development, 7*, 446–451.

Guo, S., Dan, Z., Duan, N., et al., (2018). Zn(II), Pb(II), and Cd(II) adsorption from aqueous solution by magnetic silica gel: Preparation, characterization, and adsorption. *Environmental Science and Pollution Research, 25*, 30938–30948.

He, S., Zhong, L., Duan, J., Feng, Y., Yang, B., & Yang, L., (2017). Bioremediation of wastewater by iron oxide-biochar nanocomposites loaded with photosynthetic bacteria. *Frontiers in Microbiology, 8*, 823.

Helal, M. I. D., Husein, M. E., Walaa, G., & Mostafa, E. D., (2019). Characterization of agricultural residues- based nano biochar and its efficiency in adsorption/desorption of nutrients. *International Journal of Environment, 8*, 130–141.

Hossain, M. K., Strezov, V., Chan, K. Y., Ziolkowski, A., & Nelson, P. F., (2011). Influence of pyrolysis temperature on production and nutrient properties of wastewater sludge biochar. *Journal of Environmental Management, 92*, 223–228.

Itoh, T., & Takagi, Y., (2020). Laccase-catalyzed reactions in ionic liquids for green sustainable chemistry. *ACS Sustainable Chemistry and Engineering, 9*(4), doi: 10.1021/acssuschemeng.0c07097.

Jenie, S. N. A., Kristiani, A., Kustomo, S. S., & Mansur, D., (2017). Preparation of nanobiochar as magnetic solid acid catalyst by pyrolysis-carbonization from oil palm empty fruit bunches. In: *Proceedings of the* **3rd** *International Symposium on Applied Chemistry*. https://doi.org/10.1063/1.5011875. AIP Conf. Proc.1904, 020018-1–020018-6.

Jenie, S. N. A., Kristiani, S. A., Khaerudini, D. S., & Takeishi, K., (2018). Magnetic nanobiochar as heterogeneous acid catalyst for esterification reaction. *Journal of Environmental Chemical Engineering, 8*, 103912.

Jiang, C., Bo, J., Xiao, X., Zhang, S., Wang, Z., Yan, G., Wu, Y., et al., (2020). Converting waste lignin into nano-biochar as a renewable substitute of carbon black for reinforcing styrene butadiene rubber. *Waste Management, 102*, 732–742.

Jones, D. L., Edwards-Jones, G., & Murphy, D. V., (2011). Biochar mediated alterations in herbicide break-down and leaching in soil. *Soil Biology and Biochemistry, 43*, 804–813.

Joseph, S., Graber, E. R., Chia, C., et al., (2013). Shifting paradigms: Development of high–efficiency biochar fertilizers based on nano-structures and soluble components. *Carbon Management, 4*, 323–343.

Khalid, S., Shahid, M., Murtaza, B., et al., (2019). A critical review of different factors governing the fate of pesticides in soil under biochar application. *Science of The Total Environment, 1*, 134645.

Khorram, M. S., Wang, Y., Jin, X., Fang, H., & Yu, Y., (2015). Reduced mobility of fomesafen through enhanced adsorption in biochar-amended soil. *Environmental Toxicology and Chemistry, 34*, 1258–1266.

Khorram, M. S., Zhang, Q., Lin, D., et al., (2016). Biochar: A review of its impact on pesticide behavior in soil environments and its potential applications. *Journal of Environmental Science, 44*, 269–279.

Kirchherr, J., Reike, D., & Hekkert, M., (2017). Conceptualizing the circular economy: An analysis of 114 definitions. *Resources, Conservation and Recycling, 127*, 221–232.

Kolton, M., Harel, Y. M., Pasternak, Z., Graber, E. R., Elad, Y., & Cytryn, E., (2011). Impact of biochar application to soil on the root-associated bacterial community structure of fully developed greenhouse pepper plants. *Applied and Environmental Microbiology, 14*, 4924–4930.

Kumar, M., Xiong, X., Wan, Z., Sun, Y., Tsang, D. C., Gupta, J., Gao, B., et al., (2020). Ball milling as a mechanochemical technology for fabrication of novel biochar nanomaterials. *Bioresources Technology*, 123613.

Laird, D. A., Fleming, P. D., Davis, D. D., Wang, B., Horton, R., & Karlen, D. L., (2010). Impact of biochar amendments on the quality of a typical Midwestern agricultural soil. *Geoderma, 158*, 443–449.

Lateef, A., Nazir, R., Jamil, N., Alam, S., Shah, R., Khan, M. N., & Saleem, M., (2019). Synthesis and characterization of environmental friendly corncob biochar based nano-composite–A potential slow release nano-fertilizer for sustainable agriculture. *Environmental Nanotechnology, Monitoring and Management*, 11.

Lehmann, J., Kern, D., German, L., Mccann, J., Martins, G. C., & Moreira, A., (2003). Soil fertility and production potential. In: Lehmann, J., Kern, D. C., Glaser, B., & Wodos, W. I., (eds.), *Amazonian Dark Earths* (pp. 105–124). Springer, Dordrecht.

Lehmann, J., Rillig, M. C., Thies, J., Masiello, C. A., Hockaday, W. C., & Crowley, D., (2011). Biochar effects on soil biota–A review. *Soil Biology and Biochemistry, 43*, 1812–1836.

Li, B., Yang, L., Wang, C., Zhang, Q., Liu, Q., Li, Y., & Xiao, R., (2017). Adsorption of Cd(II) from aqueous solutions by rape straw biochar derived from different modification processes. *Chemosphere, 175*, 332–340.

Li, C., Zhang, L., Gao, Y., & Li, A., (2018). Facile synthesis of nano ZnO/ZnS modified biochar by directly pyrolyzing of zinc contaminated corn stover for Pb(II), Cu(II) and Cr (VI) removals, *Waste Management, 79*, 625–637.

Li, R. H., Zhang, Y. C., Deng, H. X., Zhang, Z. Q., Wang, J. J., Shaheen, S. M., Xiao, R., et al., (2020). Removing tetracycline and Hg(II) with ball-milled magnetic nanobiochar and its potential on polluted irrigation water reclamation. *Hazard Materials, 384*, 121095.

Li, R., Liang, W., Wang, J. J., Gaston, L. A., et al., (2018). Facilitative capture of As(V), Pb(II) and methylene blue from aqueous solutions with MgO hybrid sponge-like carbonaceous composite derived from sugarcane leafy trash. *Journal of Environmental Management, 212*, 77–87.

Li, R., Zhang, Y., Deng, H., Zhang, Z., Wang, J. J., Shaheen, S. M., & Du, J., (2019). Removing tetracycline and Hg (II) with ball-milled magnetic nanobiochar and its potential on polluted irrigation water reclamation. *Journal of Hazardous Materials*, 121095.

Lian, F., Yu, W., Zhou, Q., Gu, S., Wang, Z., & Xing, B., (2020). Size matters: Nano-biochar triggers decomposition and transformation inhibition of antibiotic resistance genes in aqueous environments. *Environmental Sciences and Technology Journal, 54*, 8821–8829.

Liang, B., Lehmann, J., Solomon, D., et al., (2006). Black carbon increases cation exchange capacity in soils. *Soil Science and Society of America Journal, 70*, 1719–1730.

Liao, H., Li, Y., & Yao, H., (2019). Biochar amendment stimulates utilization of plant-derived carbon by soil bacteria in an intercropping system. *Frontiers in Microbiology, 10*.

Lin, R., Cheng, J., Ding, L., & Murphy, J. D., (2018). Improved efficiency of anaerobic digestion through direct interspecies electron transfer at mesophilic and thermophilic temperature ranges. *Chemical Engineering Journal, 350*, 681–691

Liu, G., Zheng, H., Jiang, Z., Zhao, J., Wang, Z., Pan, B., & Xing, B., (2018). Formation and physicochemical characteristics of nano biochar: Insight into chemical and colloidal stability. *Environmental Science Technology, 52*, 10369–10379.

Liu, S. B., Tan, X. F., Liu, Y. G., et al., (2016). Production of biochars from Ca impregnated ramie biomass (*Boehmeria nivea* (L.) Gaud.) and their phosphate removal potential. *RSC Advances, 6*, 5871–5880.

Liu, W., Li, Y., Feng, Y., et al., (2020). The effectiveness of nanobiochar for reducing phytotoxicity and improving soil remediation in cadmium-contaminated soil. *Scientific Reports, 10*, 858.

Liu, X., Zhang, A., Ji, C., Joseph, S., Bian, R., Li, L., Pan, G., & Paz-Ferreiro, J., (2013). Biochar's effect on crop productivity and the dependence on experimental conditions—A metaanalysis of literature data. *Plant Soil, 373*, 583–594.

Lonappan, L., Liu, Y., Rouissi, T., Pourcel, F., Brar, S. K., Verma, M., & Surampalli, R. Y., (2018). Covalent immobilization of laccase on citric acid functionalized micro-biochars derived from different feedstock and removal of diclofenac. *Chemical Engineering Journal, 351*, 985–994.

Lyu, H., Gao, B., He, F., Zimmerman, A. R., Ding, C., Huang, H., & Tang, J., (2018). Effects of ball milling on the physicochemical and sorptive properties of biochar: Experimental observations and governing mechanisms. *Environmental Pollution, 233*, 54–63.

Ma, S. Q., Jing, F. Q., Sohi, S. P., & Chen, J. W., (2019). New insights into contrasting mechanisms for PAE adsorption on millimeter, micron and nano-scale biochar. *Environmental Science and Pollution Research, 26*, 18636–18650.

Mahmoud, E., Baroudy, A. E. L., Ali, N., & Sleem, M., (2020). Spectroscopic studies on the phosphorus adsorption in salt-affected soils with or without nano-biochar additions. *Environmental Research, 184*, 109277.

Mahmoud, M. E., El-Ghanam, A. M., Saad, S. R., & Mohamed, R. H. A., (2020). Promoted removal of metformin hydrochloride anti-diabetic drug from water by fabricated and modified nanobiochar from artichoke leaves. *Sustainable Chemistry and Pharmacy, 18*, 100336.

Mangrich, A. S., Cardos, E. M. C., Doumer, M. E., Romao, L., Vidal, M., Rigol, A., & Novotny, E., (2015). Improving the water holding capacity of soils of Northeast Brazil by Biochar Augmentation. *ACS Symposium Series, 1206*, 339–354.

Manya, J. J., (2012). Pyrolysis for biochar purposes: A review to establish current knowledge gaps and research needs. *Environmental Science and Technology, 46*, 7939–7954.

Meyer, S., Glaser, B., & Quicker, P., (2011). Technical, economical, and climate-related aspects of biochar production technologies: A literature review. *Environmental Sciences and Technology, 45*, 9473–9483.

Mirbagheri, S. A., Safieddin, A. S. M., & Kiani, D. K. M., (2020). Modeling of the engine performance and exhaust emissions characteristics of a single-cylinder diesel using nano-biochar added into ethanol-biodiesel-diesel blends. *Fuel, 278*, 118238.

Nag, S. K., Kookana, R., Smith, L., Krull, E., Macdonald, L. M., & Gill, G., (2011). Poor efficacy of herbicides in biochar–amended soils as affected by their chemistry and mode of action. *Chemosphere, 84*, 1572–1577.

Naghdi, M., Taheran, M., Brar, S. K., et al., (2018). Pinewood nanobiochar: A unique carrier for the immobilization of crude laccase by covalent bonding. *International Journal of Biological Macromolecules, 115*, 563–571.

Naghdi, M., Taheran, M., Brar, S. K., Kermanshahi-pour, A., Verma, M., & Surampalli, R. Y., (2017). Immobilized laccase on oxygen functionalized nanobiochars through mineral acids treatment for removal of carbamazepine. *Science of The Total Environment, 584*, 393–401.

Naghdi, M., Taheran, M., Brar, S. K., Kermanshahi-pour, A., Verma, M., & Surampalli, R. Y., (2018). Pinewood nanobiochar: A unique carrier for the immobilization of crude laccase by covalent bonding. *International Journal of Biological Macromolecules, 115*, 563–571.

Naghdi, M., Taheran, M., Brar, S. K., Kermanshahi-Pour, A., Verma, M., & Surampalli, R. Y., (2019). Fabrication of nanobiocatalyst using encapsulated laccase onto chitosannanobiochar composite. *International Journal of Biological Macromolecules, 124*, 530–536.

Nath, B. K., Chaliha, C., & Kalita, E., (2019). Iron oxide permeated mesoporous rice-husk nanobiochar (IPMN) mediated removal of dissolved arsenic (As): Chemometric modelling and adsorption dynamics. *Journal of Environmental Management, 246*, 397–409.

Orr, C. H., & Ralebitso-Senior, T. K., (2014). Tracking N–cycling genes in biochar–supplemented ecosystems: A perspective. *OA Microbiology, 2*, 1.

Pan, F., Chapman, S. J., Li, Y., & Yao, H., (2017). Straw amendment to paddy soil stimulates denitrification but biochar amendment promotes anaerobic ammonia oxidation. *Journal of Soils Sediments, 17*, 2428–2437.

Qian, K., Kumar, A., Zhang, H., Bellmer, D., & Huhnke, R., (2015). Recent advances in utilization of biochar. *Renewable and Sustainable Energy Reviews, 42*, 1055–1064.

Qian, L., Chen, L., Joseph, S., Pan, G., Li, L., et al., (2014). Biochar compound fertilizer as an option to reach high productivity but low carbon intensity in rice agriculture of China. *Carbon Management, 5*, 145–154.

Rajamma, R., Ball, R. J., Tarelho, L. A., Allen, G. C., Labrincha, J. A., & Ferreira, V. M., (2009). Characterisation and use of biomass fly ash in cement-based materials. *Journal of Hazardous Materials, 172*, 1049–1060.

Rajput, V. D., Kumari, A., Minkina, T., Barakhov, A., Singh, S., et al., (2022a). A practical evaluation on integrated role of biochar and nanomaterials in soil remediation processes. *Environmental Geochemistry and Health*, 1–15.

Rajput, V. D., Minkina, T., Ahmed, B., Singh, V. K., Mandzhieva, S., et al., (2022b). Nano-biochar: A novel solution for sustainable agriculture and environmental remediation. *Environmental Research, 210*, 112891.

Ramadan, M. M., Amal, A., & Abd-Elsalam, K. A., (2020). Micro/nano biochar for sustainable plant health: Present status and future prospects. In: Kamel, A., & Abd-Elsalam (eds.), *Carbon Nanomaterials for Agri-Food and Environmental Applications* (pp. 323–357). Elsevier BV.

Ramanayaka, S., Kumar, M., Etampawala, T., & Vithanage, M., (2020). Macro, colloidal and nanobiochar for oxytetracycline removal in synthetic hydrolyzed human urine. *Environmental Pollution, 267*, 115683.

Ramanayaka, S., Tsang, D. C. W., Hou, D., et al., (2019). Green synthesis of graphitic nanobiochar for the removal of emerging contaminants in aqueous media. *Science of The Total Environment, 310*, 107306.

Ramanayaka, S., Vithanage, M., Alessi, D. S., Liu, W. J., Jayasundera, A. C. A., & Ok, Y. S., (2020). Nanobiochar: Production, properties, and multifunctional applications. *Environmental Science Nano, 7*, 3279–3302.

Rawat, J., Saxena, J., & Sanwal, P., (2019). In: Abrol, V., & Sharma, P., (eds.), *Biochar: A Sustainable Approach for Improving Plant Growth and Soil Properties, Biochar – an Imperative Amendment for Soil and the Environment.* IntechOpen. https://doi.org/10.5772/intechopen.82151.

Reddy, D. H. K., & Lee, S. M., (2014). Magnetic biochar composite: Facile synthesis, characterization and application for heavy metal removal. *Colloids Surfaces A: Physicochemical and Engineering Aspects, 454*, 96–103.
Rondon, M. A., Lehmann, J., Ramírez, J., & Hurtado, M., (2007). Biological nitrogen fixation by common beans (*Phaseolus vulgaris* L.) increases with bio–char additions. *Biology and Fertility of Soils, 43*, 699–708.
Rose, P., Bhat, M., Vidhani, K., Ajmeri, N., Gole, A., & Ghaisas, S., (2011). Intelligent informatics platform for nano–agriculture. In: *11***th** *IEEE International Conference on Nanotechnology* (pp. 916–919). IEEE.
Rossow, M., (2003). *Fly Ash Facts for Highway Engineers*. New York NY Continuing Education and Development.
Safieddin, A. S. M., Taghipoor, A., Solmaz, H., & Mostafaei, M., (2020). The effect of nano-biochar on the performance and emissions of a diesel engine fueled with fuel oil-diesel fuel. *Fuel, 268*, 117356.
Samsudin, M. H., Hassan, M. A., Idris, J., et al., (2019). A one-step self-sustained low temperature carbonization of coconut shell biomass produced a high specific surface area biochar derived nano-adsorbent. *Waste Management Research, 37*, 551–555.
Saxena, J., Rawat, J., & Sanwal, P., (2016). Enhancement of growth and yield of glycine max plants with inoculation of phosphate solubilizing fungus *Aspergillus Niger* K7 and biochar amendment in soil. *Commun. Soil Science and Plant Analysis, 47*, 2334–2347.
Saxena, M., Maity, S., & Sarkar, S., (2014). Carbon nanoparticles in 'biochar' boost wheat (*Triticum aestivum*) plant growth. *RSC. Advances, 4*, 39948–39954.
Senthilkumar, R., & Prasad, D. M. R., (2020). In: Ahmed, A. A., & Mohammed, H. H. A., (eds.), *Sorption of Heavy Metals onto Biochar, Applications of Biochar for Environmental Safety.* IntechOpen. https://www.intechopen.com/chapters/72272 (accessed on 25 April 2024).
Sigmund, G., Jiang, C., Hofmann, T., & Chen, W., (2018). Environmental transformation of natural and engineered carbon nanoparticles and implications for the fate of organic contaminants. *Environmental Science Nano, 5*, 2500–2518.
Soja, G., Zehetner, F., & Kitzler, B., (2013). *Biochar as Yield Promoter for Agricultural Crops? Conjectures and Observations*. Available from: http://www.oeaw.ac.at/forebiom/WS1lectures/SessionIII_Soja.pdf (accessed on 25 April 2024).
Song, B., Lin, R., Lam, C. H., Wu, H., Tsui, T. H., & Yu, Y., (2021). Recent advances and challenges of inter-disciplinary biomass valorization by integrating hydrothermal and biological techniques. *Renewable Sustainable Energy Review, 135*, 110370.
Taheran, M., Naghdi, M., Brar, S. K., Verma, M., & Surampalli, R. Y., (2018). Emerging contaminants: Here today, there tomorrow. *Environmental Nanotechnology and Monitoring Management, 10*, 122–126.
Tan, X. F., Liu, Y. G., Gu, Y. L., et al., (2016). Biochar-based nano-composites for the decontamination of wastewater: A review. *Bioresources Technology, 212*, 318–333.
Tan, X. F., Liu, Y. G., Zeng, G., et al., (2015). Application of biochar for the removal of pollutants from aqueous solutions. *Chemosphere, 125*, 70–85.
Thies, J. E., Rillig, M. C., & Graber, E. R., (2015). Biochar effects on the abundance, activity and diversity of the soil biota, In: Lehmann, J., & Joseph, S., (eds.), *Biochar for Environmental Management: Science, Technology and Implementation* (2nd edn., pp. 327–389). Earthscan Publications Ltd., London, UK.

Timmons, D., Lema-Driscoll, A., & Gazi, U. M. A., (2017). *The Economics of Biochar Carbon Sequestration in Massachusetts*. Available from: https://ag.umass.edu/sites/ag.umass.edu/files/reports/timmons_-_biochar_report_10-16-17.pdf (accessed on 25 April 2024).

Tiwari, D. K., Dasgupta-Schubert, N., Cendejas, L. V., Villegas, J., Montoya, L. C., & García, S. B., (2014). Interfacing carbon nanotubes (CNT) with plants: Enhancement of growth, water and ionic nutrient uptake in maize (*Zea mays*) and implications for nanoagriculture. *Applied Nanosciences, 4*, 577–591.

Ubando, A. T., Felix, C. B., & Chen, W. H., (2020). Biorefineries in circular bioeconomy: A comprehensive review. *Bioresources Technology, 299*, 122585.

Verheijen, F., Jeffery, S., Bastos, A., Van, D. V. M., & Diafas, I., (2010). *Biochar Application to Soils: A Critical Scientific Review of Effects on Soil Properties, Processes, and Functions*. EUR 24099: 162.

Wan, X., Li, C., & Parikh, S. J., (2020). Simultaneous removal of arsenic, cadmium and lead from soil by iron-modified magnetic biochar. *Environmental Pollution, 261*, 114157.

Wang, J., & Wang, S., (2019). Preparation, modification and environmental application of biochar: A review. *Journal of Cleaner Production, 227*, 1002–1022.

Wang, S., Dai, G., Yang, H., & Luo, Z., (2017). Lignocellulosic biomass pyrolysis mechanism: A state of- the-art review. *Progress in Energy and Combustion Science, 62*, 33–86.

Wang, X., Gu, Y., Tan, X., et al., (2019). Functionalized biochar/clay composites for reducing the bioavailable fraction of arsenic and cadmium in river sediment. *Environmental Toxicology and Chemistry, 38*, 2337–2347.

Weber, K., & Quicker, P., (2018). Properties of biochar. *Fuel, 217*, 240–261.

Wei, X., Wang, X., Gao, B., Zou, W., & Dong, L., (2020). Facile ball-milling synthesis of cuo/ biochar nanocomposites for efficient removal of reactive red 120. *ACS Omega, 5*, 5748–5755.

Woolf, D., Amonette, J. E., Street-Perrott, F. A., Lehmann, J., & Joseph, S., (2010). Sustainable biochar to mitigate global climate change. *Nature Communication, 1*, 1–9.

Wu, J., Li, Z., Huang, D., et al., (2020). A novel calcium-based magnetic biochar is effective in stabilization of arsenic and cadmium co-contamination in aerobic soils. *Journal of Hazard Mater, 387*, 122010.

Xiu, S., Shahbazi, A., & Li, R., (2017). Characterization, modification and application of biochar for energy storage and catalysis: A review. *Trends Renewable Energy, 3*, 86–101.

Xu, G., Lv, Y., Sun, J., Shao, H., & Wei, L., (2012). Recent advances in biochar applications in agricultural soils: Benefits and environmental implications. *Clean, 40*, 1093–1098.

Xu, R., Tang, R., Zhou, Q., Li, F., & Zhang, B., (2015). Enhancement of catalytic activity of immobilized laccase for diclofenac biodegradation by carbon nanotubes. *Chemical Engineering Journal, 262*, 88–95.

Yao, L., He, L., Yang, Y., Zhang, Y., Liu, Z., Liang, L., & Piao, Y., (2021). Nanobiochar paper based electrochemical immunosensor for fast and ultrasensitive detection of microcystin-LR. *Science Total Environment, 750*, 141692.

Yao, Y., Gao, B., Chen, J., & Yang, L., (2013). Engineered biochar reclaiming phosphate from aqueous solutions: Mechanisms and potential application as a slow release fertilizer. *Environmental Science Technology, 47*, 8700–8708.

Yao, Y., Gao, B., Inyang, M., et al., (2011). Biochar derived from anaerobically digested sugar beet tailings: Characterization and phosphate removal potential. *Bioresources Technology, 102*, 6273–6278.

Yue, L., Lian, F., Han, Y., Bao, Q., Wang, Z., & Xing, B., (2019). The effect of biochar nanoparticles on rice plant growth and the uptake of heavy metals: Implications for agronomic benefits and potential risk. *Science Total Environment, 656*, 9–18.

Zeng, Z., Zhang, S. D., Li, T. Q., et al., (2013). Sorption of ammonium and phosphate from aqueous solution by biochar derived from phytoremediation plants. *J. Zhejiang Univ. – Sci. B., 14*, 1152–1161.

Zhang, M., Gao, B., Varnoosfaderani, S., Hebard, A., Yao, Y., & Inyang, M., (2013). Preparation and characterization of a novel magnetic biochar for arsenic removal. *Bioresources Technology, 130*, 457–462.

Zhang, P., Huang, P., Xu, X., Sun, H., Jiang, B., & Liao, Y., (2020). Spectroscopic and molecular characterization of biochar-derived dissolved organic matter and the associations with soil microbial responses. *Science of The Total Environment, 708*, 134619.

Zhang, Y., Piao, M., He, L., Yao, L., Piao, T., Liu, Z., & Piao, Y., (2020). Immobilization of laccase on magnetically separable biochar for highly efficient removal of bisphenol A in water. *RSC Advances, 10*, 4795–4804.

Zhou, L., Huang, Y., Qiu, W., Sun, Z., Liu, Z., & Song, Z., (2017). Adsorption properties of nano- MnO_2–biochar composites for copper in aqueous solution. *Molecules, 22*, 173–186.

Zhu, S., Huang, X., Wang, D., Wang, L., & Ma, F., (2018). Enhanced hexavalent chromium removal performance and stabilization by magnetic iron nanoparticles assisted biochar in aqueous solution: Mechanisms and application potential. *Chemosphere, 207*, 50–59.

Zhu, Y., Hu, J., & Wang, J., (2012). Competitive adsorption of Pb (II), Cu (II) and Zn (II) onto xanthate-modified magnetic chitosan. *Journal of Hazardous Materials, 221*, 155–161.

CHAPTER 3

Nano-Biochar: A Soil Conditioner for Sustainable Soil Health

P. KANNAN,[1] M. MOHAMED ROSHAN ABU FIRNASS,[1] S. PONMANI,[2] D. KRISHNA VENI,[3] and C. SWAMINATHAN[4]

[1]*Department of Soils and Environment, Agricultural College and Research Institute, Madurai, Tamil Nadu, India*

[2]*Department of Environmental Sciences, Mother Terasa College of Agriculture, Pudukkottai, Tamil Nadu, India*

[3]*Department of Chemistry, Alagappa Government Arts College, Karaikudi, Tamil Nadu, India*

[4]*Water Technology Center, Tamil Nadu Agricultural University, Coimbatore, Tamil Nadu, India*

ABSTRACT

Carbon nanomaterials have emerged as strong tools in recent years as a result of their unique properties and wide range of applications in fields such as agriculture and the environment. Biochar technology and nanotechnologies have the potential to produce carbon-based nanomaterials such as nano-biochar owing to their varied characteristics. Nano-biochar is a nanosized biochar substance with improved physical, chemical, and surface characteristics. It offers various advantages, including improved crop growth and soil characteristics, bioremediation of pollutants, wastewater treatment, carbon storage, and enzyme immobilization. Because of its economic effectiveness, sustainability, and eco-friendliness, it may become the ideal alternative to traditional techniques. It can also help to prevent climate change through its carbon sequestration capability. The application of nano-biochar as a

Nanomaterials and Nano-Biochar in Reducing Soil Stress: An Integrated Approach to Sustainable Agriculture. Vishnu D. Rajput, & Tatiana Minkina (Eds.)

soil amendment enhanced soil pH, EC, CEC, organic matter, soil porosity, infiltration rate, and microbial activity. In comparison to biochar, nano-biochar absorbs pollutants and nutrients well and is mobile in soil, making it a possible option for waste management. However, data on the function of nano-biochar in altering soil quality is currently lacking. This chapter gives us a clear picture of how nano-biochar may be utilized to improve the physio-chemical qualities of soil.

3.1 INTRODUCTION

The rapid growth of population and high rates of urbanization have been increasing, threatening global food security and farming regions. Immediate changes are necessary to enhance the existing depleted farmed soil regions in an environmentally appropriate manner. Enhanced population necessitates additional agricultural goods, leading to increased agricultural residual output (Wrobel-Tobiszewska et al., 2014). To meet crop productivity demands, chemical fertilizers have been used for centuries to boost crop production and soil quality. However, recent studies indicate that these fertilizers have a negative rather than a positive impact on the environment (Xu et al., 2018). As an alternative to chemical fertilizers, a greater number of slow-release fertilizers have been developed, notably those bound with various types of polymers or encapsulated with zeolite, etc. (Li et al., 2017a; Yuvaraj & Subramanian, 2018). Among them, biochar-based fertilizer is an effective slow-release fertilizer that reduces unnecessary nutrient losses from the soil ecosystem. It also improves soil fertility, promotes plant growth, and increases crop yield (Yuan et al., 2017).

The notion of biochar originates from the ancient Amerindian cultures of the Amazon region of South America (Woods et al., 2009). Studies on Amazonian black earth, the human-made soil combination called terra preta, suggest that these early civilizations used a carbonaceous material comparable to biochar to boost soil fertility and agricultural productivity (Many, 2012; Ahmad et al., 2014). Similar carbon-rich substances have been used by humans for ages; however, experts coined the term "biochar" to indicate a pyrogenic carbon-containing solid material developed in an oxygen-free environment at extreme temperatures, which was earlier branded "agrichar" (Nath et al., 2019). Biochar is frequently utilized in a variety of application areas, including management of solid waste, soil conditioning, water purification, and carbon capture and storage (Ahmad et al., 2014). Biochar's porous structure, functional groups, and surface area all aid in immobilizing a variety of contaminants, as well as highly toxic metals (Cao & Harris, 2010). Pyrolysis converts waste into high-carbon biochar, which is extensively used as a low-cost and economical product due to its distinct structural, physical, and chemical features. It can sequester carbon (Sohi et al., 2010), reduce N_2O emissions, and enrich soil habitats (Blanco-Canqui, 2021).

Furthermore, they increase microbial population in the soil and increase fungal utilization of carbon matter in the soil while maintaining chemical and biological features (Liu et al., 2020a; Meng et al., 2019). Biochar can also regulate root growth and root structure and boost the photosynthesis rate in crops (He et al., 2020; Liu et al., 2021a). Recent research has focused on the utilization of nano-biochar to enhance soil performance and quality for enhanced crop growth and yield quality (Yang et al., 2021; Liu et al., 2021b; Wu et al., 2019; Wang et al., 2017a) by acting as media to provide mechanical support for microbes (Duan et al., 2021; An et al., 2021) and possibly enhance soil physico-chemical characteristics by governing soil pH (Yang et al., 2021), electron conductivity (Zhang et al., 2021), cation exchange capacity (Duan et al., 2021), organic matter (Moradi & Karimi, 2021), and soil aggregation (Duan et al., 2021; An et al., 2021).

With the evolution of nanotechnology, studies on the sustainable creation of nano-biochar for soil and agricultural applications have been conducted (Gao & Wu, 2014). In this chapter, we refer to biochars with particle sizes of 1 μm or larger as macro biochar, 1 μm-100 nm as colloidal biochar, and 100 nm as nano biochar (Liu et al., 2018). Nano biochar has a significantly greater surface area to mass ratio than macro biochar, and it has several possible applications other than sorbent, such as a capacitor, sensor, and photocatalytic material (Spokas et al., 2014). The most frequent method for producing nanobiochar is to scale down biochar to the nanoscale via a top-down method. Nanobiochar's physico-chemical and mechanical qualities are critical in deciding its usage. Biochar yield, chemical compositions, surface area, and porosity are all influenced by feedstock qualities, the manufacturing method, temperature, and residence time (Weber & Quicker, 2018). However, pre- and post-modification procedures can be used to improve the qualities of biochar (Azargohar & Dalai, 2008).

Due to its high specific surface area and micro-porosity, nano-biochar has a high adsorption potential for a wide range of pollutants, including potentially hazardous metals, pesticides, hydrocarbons (PAHs), and can neutralize low pH soils owing to its extremely alkaline nature (Wang et al., 2013; G'amiz et al., 2017). Mechanical grinding results in a new technique for biochar production in which the material may be tailored by reducing particle size to the nano-scale. This subject has just emerged as a novel area of study. Mechanical crushing is the thoroughly prevalent method for producing nano-biochar (Naghdi et al., 2017a). Given the recent surge in interest, the vital goal of this section is to reveal the soil physicochemical characteristics of nano-biochar and their relationship with the growth and development of crops which stated, thus far in an attempt to offer a thoughtful evaluation of knowledge gaps, thus fully examine suggestions for future research.

3.2 PRODUCTION OF NANO-BIOCHAR

Biochar is generally produced through a process called pyrolysis, in which feedstock is thermally decomposed by heating it between 350–700°C in an environment with reduced oxygen (<1%) (Oleszczuk et al., 2016). When lignocellulosic biomass, such as plant residues, wood biomass, stable waste, and animal waste, undergoes various thermochemical processes, such as rapid or slow pyrolysis, torrefaction, carbonization using hydrothermal or flash, and gasification approaches, biochar is produced (Figure 3.1) (Meyer et al., 2011). Biochar is typically produced through slow pyrolysis in an oxygen-free environment, with a characteristic dry biomass yield of 35% and a small amount of bio-oil (Mohan et al., 2014). While fast pyrolysis is preferred over other methods for producing biofuels, gasification is mostly used to produce syngas, which is then used to generate heat and energy. It has also been shown that lignocellulosic materials produce more biochar than municipal solid waste. Additionally, biochar is physically and chemically modified for various purposes to enhance its functionality. Chemical oxidation, CO_2 activation, steam coating, acid/base treatment, impregnation by natural and synthetic nanoparticles, etc. are used in chemical modification treatments. In contrast, physical modification techniques like ball milling have not gained much popularity, even though the limited research in this area has shown promising results that require further investigation.

Currently, nanobiochar is created from regular biochar using several methods, as revealed in Figure 3.1. Typically, mechanical crushing is done to create nanosize particles. Direct nanobiochar production through flash heating produced graphitic nanosheets despite mechanical processes (Ramanayaka et al., 2020). To physically scatter the biochar, Oleszczuk et al. (2016) applied an ultrasonic vibrator. Sonication was then used to create nanosize particles. The most acceptable and preferable technique of producing biochar, according to the literature that is currently accessible, is through the ball milling process. The process of grinding in a ball mill involves the collision of particles that agglomerate to ostensibly enhance the particle size. The creation of nanobiochar using double-disc milling may also be an alternative; however, it has not been used very frequently due to its greater operating costs. It has been found that the vibrating disc milling approach outperforms the most popular ball mill method in terms of producing a greater quantity of nanobiochar with improved consistency of size and shape by attrition and shear stress. Guo et al. (2020) found that a hydrothermal reaction employing farming by-products as well as residues was another way to produce nanobiochar.

Bulk biochar was produced using bovine manure and soybean straw as the feedstock. Then, it was broken down in a hydrothermal reactor at high pressure using concentrated acids (sulfuric acid and nitric acid). Following digestion, the suspension was filtered through a 0.22 mm membrane and then heated to 65°C for drying. Using this technique, 5-nm and 4-nm-sized biochar nanodots were created. There are currently several ways to create nanobiochar, each with advantages and limitations of their own. Various nanobiochar types, raw materials, and applications with various temperature levels are depicted in Table 3.1.

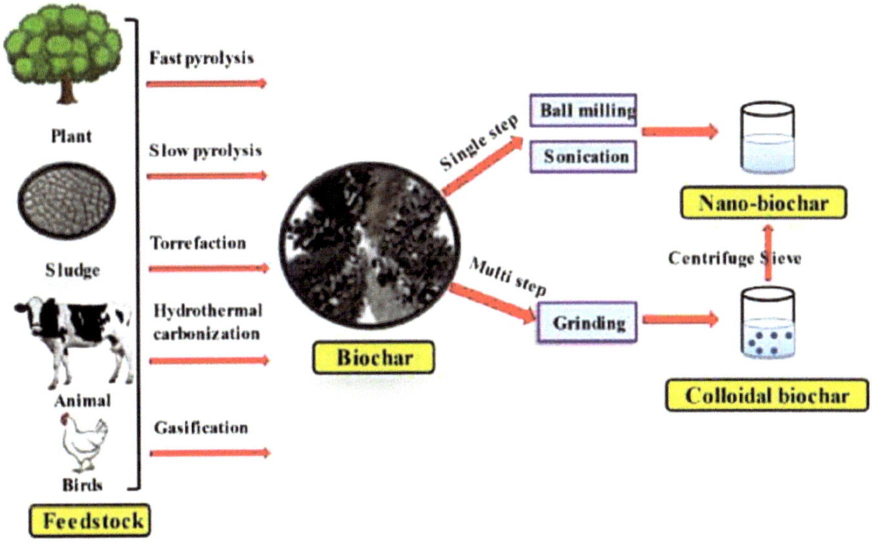

FIGURE 3.1 Production methods of nano-biochar

TABLE 3.1 Numerous types of nanobiochar, feedstock materials used, and their applications at various temperatures

Nano-biochar type	Feedstock	Application	Temperature (°C)	References
Bagasse biochar	Sugarcane bagasse	Sulfamethoxazole and sulfa-pyridine antibiotic removal from aqueous systems	450°C	Huang et al. (2020)
		Carbon electrode as an alternative	600°C	Lyu et al. (2019)
Bamboo biochar	Bamboo	Ammonium removal from water	300°C	Qin et al. (2020)
		Aqueous methylene blue removal	450°C	Lyu et al. (2018)
Wood biochar	Pinewood	Carrier for crude laccase immobilization via covalent bonding	525°C	Naghdi et al. (2018)
		Removing contaminants (carbamazepine)	525°C	Naghdi et al. (2019)
Rice husk nano-biochar	Rice husk	Removal of metal ion [As (III)]	600°C	Nath et al. (2019)
		Phthalate esters adsorption	500°C	Ma et al. (2019)
Sorghum straw biochar	Sorghum straw	Carbonaceous nanomaterials derived from biochar for heavy metal sensing in aqueous environments	500°C	Plácido et al. (2019)

3.3 PROPERTIES OF NANO-BIOCHAR

According to reports, nano-biochar has properties similar to carbon-based products (Laird et al., 2010; Chen et al., 2011). The synthesis technique and feedstock materials used substantially influence the properties of nano-biochar. An FTIR study revealed that lignin-derived nano-biochar contains several functional groups, including hydroxyl, carboxylic, carbonyl, and aromatic characteristics (Jiang et al., 2020; Naghdi et al., 2017a). Furthermore, changes in particle size, elemental content (C, H, N, S), electrical conductivity, and X-ray diffraction patterns may be observed in manufactured nano-biochar (Mahmoud et al., 2020; Naghdi et al., 2017b, 2018; Ardebili et al., 2020). As a result, a complete physicochemical characterization of nano-biochar would aid in determining its applicability for the required purposes. The diverse properties of biochar allow for multiple uses in various industries. Nano-biochar, which is made by reducing the particle size of biochar, may have properties comparable to macro-biochar.

3.3.1 PHYSICAL PROPERTIES OF BIOCHAR

3.3.1.1 SURFACE AREA

Biochar has a large surface area due to the enormous amount of pores found in the substance (Brewer et al., 2014). The overall surface area is directly influenced by the number of volatile gases emitted during the carbonization method. However, hydrothermal carbonization diminishes the surface area, while pyrolysis produces biochar with a large surface area (Weber & Quicker, 2018). In theory, nanoparticles should have relatively large surface areas as compared to particles at the macro size. Significant changes in the surface area have been observed at various pyrolysis temperatures. Furthermore, nano-biochars generated at low temperatures, i.e., 300–400°C, have modest surface areas ranging from 5.6 to 47.2 m^2g^{-1}, but high temperatures, such as 450 and 600°C, create a substantially greater surface area range (342 to 430 m^2g^{-1}) (Xu et al., 2019; Huang et al., 2020; Li et al., 2020). The rise in surface area is expected due to the thermal degradation of biomass substances and the formation of pores on the material's surface.

3.3.1.2 WATER RETENTION CAPACITY

The hydrologic characteristics of nano-biochar are directly dependent on the pyrolysis temperature and the porosity of the feedstock. The number of surface functional groups decreases as pyrolysis temperatures rise, resulting in lesser water retention capacity and higher hydrophobicity (Alam et al., 2018a, b; Weber & Quicker, 2018). As a result, no significant variation in water retention ability is seen through decreasing particle size depending on pyrolysis temperature. However, the capacity of the substance to hold water relies on the porosity of the material. Nano-biochars manufactured by reduced pyrolytic temperatures also have pores that are too tiny to retain a significant quantity of water (Das & Sarmah, 2015).

3.3.1.3 ZETA POTENTIAL

The zeta potential of nano-biochar has been examined to better comprehend their aggregation, electrokinetics, and sustainability (Song et al., 2019; Safari et al., 2019; Yi et al., 2015). The zeta potential of nano-biochar is influenced by particle size and feedstock nature (Song et al., 2019). As a result, nano-biochar has lesser stabilities than micro or macro-biochars. This behavior was seen in a zeta potential investigation conducted by Song et al. (2019), which indicated reduced negativity in the zeta potential values of nano-biochars (1.73–13.4 mV). Moreover, the elemental composition of the feedstock, like the carboxylic and phenolic group levels, affects the zeta potential (Qian et al., 2016). This outcome was agreed upon by data from Song et al. (2019), who discovered that nano-biochars derived from plants have higher zeta potentials than nano-biochars derived from municipal sources.

3.3.2 CHEMICAL PROPERTIES

3.3.2.1 ELEMENTAL COMPOSITION

In the generation of nano-biochar, the chemical content of raw biomass is predominantly converted into carbon. Carbon concentration appears to decrease with reducing particle size, from macro to nanoscale. Carbon concentrations of wood chip biochars at the macro, micro, and nano sizes were 81.2%, 63%, and 17.7%, respectively (Song et al., 2019). Due to the separation of functional groups, the reduction in oxygen and hydrogen levels, the carbon content increases (Weber & Quicker, 2018). The content of the carbon is related to the pyrolysis temperature and feedstock carbon concentration.

3.3.2.2 STRUCTURAL COMPOSITION

Crop-derived biomass is high in cellulose and hemicellulose, as well as lignin, that aids in structural strength. Because of the partial and complete decomposition of hemicellulose, cellulose, and lignin, most of the characteristics of biomass vary dramatically when the pyrolysis temperature fluctuates between 250 and 350°C. Wallace et al. (2019) published FTIR spectra of softwood nano-biochar, that supports the existence of cellulose, hemicellulose, as well as lignin (Wallace et al., 2019). The feedstock sources have a direct influence on the mineral content of nano-biochar. Plant biochars, like wood and farming residues, have a dominating SiO_2 peak, while sludge, municipal, and livestock waste biochars have $CaCO_3$, $Ca_3(PO_4)$, and aluminosilicate mineral peaks.

3.3.2.3 pH

The technique of manufacturing generates large variations in the pH of the final nano-biochar between pyrolysis and hydrothermal carbonization (Weber & Quicker, 2018). Likewise, feedstock type has a significant influence on nano-biochar alkalinity, which may be acidified and its pH decreased by an acid-activation phase (Vithanage et al., 2015). Because nano-biochar is utilized as a soil supplement, its pH is critical for agricultural uses. Pinewood nano-biochar, for example, has a slightly alkaline pH of 6.6 owing to its higher content of cellulose and lignin (Naghdi et al., 2019). Nonetheless, elephant grass and wheat straw nano-biochars had significantly alkaline pH values of 7.8, 8.9, and 9.1, while having relatively modest concentrations of cellulose and lignin (Oleszczuk et al., 2016). Observations demonstrate that the pH of oak nano-biochar fell from 9.4 to 7.4 in one year, whereas the pH of corn fodder nano-biochar reduced from 7.6 to 1.8 within the same period. The functional groups in biochar have been ascribed to this rise and reduction (Rhoades et al., 2017; Wang et al., 2017b). Biochar has a positive or negative net charge and primarily has a lesser cation exchange capacity than soil organic material (Alameri, 2021; Saied et al., 2021; Saffari et al., 2015).

3.4 NANO-BIOCHAR APPLICATION TO SOIL

Biochar, an organic substance recognized for numerous applications in agricultural practices (Figure 3.2) such as water-holding capacity, organic carbon, fertilization, plant growth, soil amendment, and also enhancing the growth of the microbial population. Furthermore, biochar has been described to increase crop productivity, soil fertility, and plant biomass. The use of nano-biochar in agriculture has also been initiated to reduce the use of conventional fertilizers, especially nitrogen-based fertilizers, which in turn reduces nitrogen leaching into groundwater.

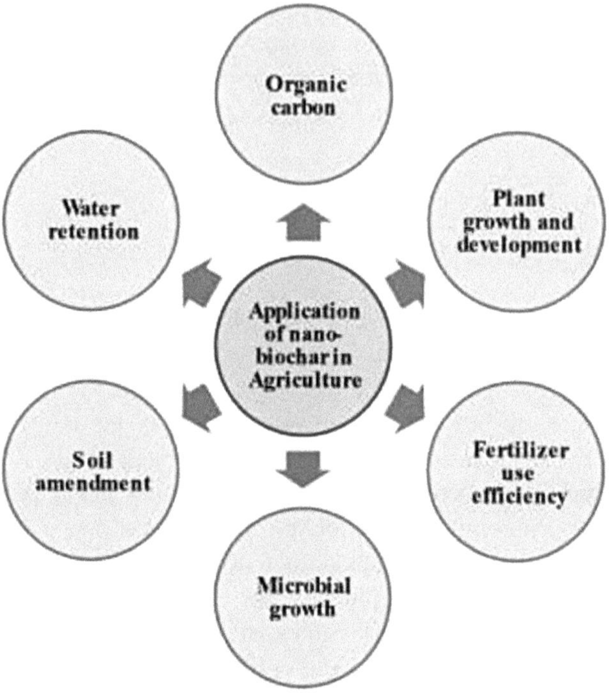

FIGURE 3.2 Application of nano-biochar to soil.

3.3.3 NANO-BIOCHAR IN SOIL WATER CHARACTERISTIC CURVES

At equivalent soil-water suction, the moisture content in soil with higher levels of nano-biochar was greater than in soil with lower levels of nano-biochar and the control (Han et al., 2010; Jun et al., 2014). This is due to the fact that the application of nano-biochar reduces the amount of macro pores while increasing the amount of micropores in the soil. Once suction is exerted, the water in the micropores is less influenced than the water in the macro pores. In addition, nano-biochar has a high specific surface area and higher water-holding capacity. As a result, nano-biochar can enhance root water uptake and improve sandy soil water retention.

3.3.4 INFLUENCE OF NANO-BIOCHAR ON SOIL NITRATE

Providing enough nitrogen fertilizer to crops may enhance nitrogen absorption and usage, boosting crop biomass and production. Honarvar et al. (2021) said that the horizontal and vertical arrangement of roots in spring maize seedlings are tiny, the root hairs are fine, and the root volume is poor. These root features hinder maize's ability to absorb nitrogen from the soil. The change in soil nitrate values before and after seeding was minimal, and the nitrate concentration was high. Some N in the soil was taken by roots of maize, while others were transferred to deep layers of soil. These variables contributed to a reduction in soil nitrogen quantity and larger nitrate variances among the original and post-sowing observations.

The variations in nitrate in the soil profile reduced as the nano-biochar concentration rose during the advanced phases of spring maize development. This showed that the presence of nano-biochar in a consistent crop reduces the amount of nitrogen delivery to the deeper layer of soil (Albiter et al., 2010). Excess N application raises the danger of nitrate leaching (Zhang et al., 2010). Dempster et al. (2012) discovered that applying nano-biochar enhanced nitrate adsorption. Applying nano-biochar at 5–10 cm under surface may reduce nitrate leaching and improve N use efficiency in soils.

3.3.5 INFLUENCE OF NANO-BIOCHAR ON SOIL AVAILABLE PHOSPHORUS

The available soil phosphorus levels of every treatment were lower after seeding than during the initial observations (Honarvar et al., 2021). The variations in soil available P among the initial and post-sowing readings in nano-biochar enriched soil were less than that of the control group in the 0–5 cm soil depth. With increasing nano-biochar concentrations, the variance of the discrepancies between the earlier and after-planting soil concentrations reduced. The variance in available soil P at 10–35 cm soil depth was minimal, while the variations in soil available phosphorus initial and after planting measures in every treatment varied substantially. This implies that the capacity of soil to retain the available P in the surface layers was enhanced when the soil was amended with nano-biochar (Sharma et al., 2015; Khorassani et al., 2011; Arif et al., 2017). Because the phosphorus was absorbed by soil particles, a significant amount of P was retained in the root zone where nano-biochar was applied. This increased the amount of P available to spring maize in the root zone.

3.3.6 INFLUENCE OF NANO-BIOCHAR ON SOIL AVAILABLE POTASSIUM

Honarvar et al. (2021) observed that the quantity of available soil K in nano-biochar supplemented soil was greater than in the control at 0–10 cm soil depth. During the tasseling stage, the average available K level in soil supplemented by nano-biochar at a range of 0.7% and 1% rose by 11.1% and 22.6%, respectively, compared to the control. When nano-biochar was mixed with manure, the plant-available K, P, Ca, and Na levels rose by a factor of 2.2, 2.5, 1.2, and 2.8, respectively (Liu et al., 2012). Nano-biochar can increase soil fertility by improving the availability of potassium in the soil (Rens et al., 2018; Dong et al., 2016). The aggregation of nano-biochars in soil enhanced the amount of micropores, preventing longitudinal transport of potassium ions.

3.3.7 INFLUENCE OF NANO-BIOCHAR ON SOIL ORGANIC CARBON CONTENT

As previously stated, soil organic carbon (SOC) breaks down over time and emits CO_2 into the surrounding environment. According to the findings, introduction of nano-biochar to the soil greatly enhances the organic carbon content (Kianfar, 2021; Syah et al., 2021; Majdi et al., 2021). For example, applying 8% nano-biochar to the soil increases the quantity of organic carbon by roughly 41% (Aramesh & Asadi, 2021). According to the findings, nano-biochar enhances the stability of soil organic carbon and decreases mineralization (Salim et al., 2021).

3.3.8 INFLUENCE OF NANO-BIOCHAR ON CO_2 EMISSIONS FROM SOIL

Initially, applying nano-biochar to the soil increases CO_2 emissions. For instance, applying 8% nano-biochar to soil boosts CO_2 releases during the first 20 days; however, gas emissions are substantially lower after 120 days (Lu et al., 2017; Luo et al., 2017a, b). Nano-biochar micrometric porosity enhances soil water retention capacity, and more moisture equals less soil organic carbon breakdown. As a result, nano-biochar indirectly decreases CO_2 emissions by increasing soil moisture (Atkinson et al., 2010).

3.3.9 INFLUENCE OF NANO-BIOCHAR ON SOIL MICROBES

The application of nano-biochar into soil may have a significant influence on soil characteristics, particularly biological ones that are mostly driven by a varied variety of microbial communities (Lian et al., 2018). The interactions of nano-biochar with soil minerals are discussed to increase the mobilization of macro-nutrients such as P (Mahmoud et al., 2020). Such changes in soil nutrients might be a deciding factor in the number of soil microorganisms

(Qu et al., 2016). Moreover, variations in the quantity of soil-applied nano-biochar resulted in changes in the prevalence of microbial populations pertaining to a certain category (Feng et al., 2021). The change in microbial community features may be attributed to changes in soil properties like water retention ability, porosity, and tensile strength following nano-biochar addition to the soil (Liu et al., 2021a).

3.3.10 INFLUENCE OF NANO-BIOCHAR ON SEED GERMINATION AND PLANT GROWTH

To investigate the impact of nano-biochar on germination, Liu et al. (2020b) conducted an experiment using four vegetable seedlings subjected to varying quantities of nano-biochar suspension. When cultivated on 500 mg/L nano-biochar treated filter paper, the germination percentages rose by 11%, 10%, 17.5%, and 10.3%, respectively, when compared to the control ($P < 0.05$). *Brassica chinensis* L. was used to observe how nano-biochar affected plant development. Measurements were taken at 15 days, 30 days, 45 days, 60 days, and 90 days after planting (Liu et al., 2018). The fresh weight and height of *Brassica* were considerably elevated due to the application of nano-biochar to Cd-polluted soil (all $P < 0.05$). Increased levels of nano-biochar did not enhance dry weight much at the start of the trial (before 45 days); however, the dry weight rose significantly at 60 days under the 0.5% nano-biochar application ($P < 0.05$).

3.3.11 INFLUENCE OF NANO-BIOCHAR ON SOIL-PLANT RELATIONSHIP

Plant roots and other vital nutrients readily absorb nano-biochar suspended or scattered in water (Liu et al., 2020a). Rice grew well, and seedling morphology improved as a result of the combined effect of nano-biochar and ferulic acid (FA). The chlorophyll content indicated normalization and restoration at percentages of 55% and 96%, respectively, after treatments with 400 mg/L and 400 mg/kg of nano-biochar combined with FA in hydroponic conditions and pot treatments (Shen et al., 2020). The input of nano-biochar into the soil system has the potential to reduce metal toxicity. Liu et al. (2020a) showed decreased bioavailability of Cd levels in soil modified with nano-biochar. The use of nano-biochar increased microbial biomass and the richness of certain bacterial communities. Wang et al. (2016) discovered that fertilizer containing nano-biochar potentially boosts crop productivity by up to 10% to 20% without lowering fertilizer requirements by 30% to 50%.

3.4 APPLICATIONS OF NANO-BIOCHAR

Nano-biochar has been employed in various activities, such as pollution removal, catalysis, biomolecule transportation, sensing materials, and components for the battery industry (Lyu et al., 2018; Naghdi et al., 2019; Jiang et al., 2020; Banek et al., 2018). Furthermore, nano-biochar can be used to enhance the growth of crops like rice and wheat (Saxena et al., 2014). The majority of nano-biochar research papers focus on its use as an adsorbent material. The capacity of nano-biochar to remove various contaminants, including organic and inorganic molecules, potentially harmful components, agrochemicals, and medications, has been investigated (Wallace et al., 2019).

However, apart from ball-milling for nano-biochar as a direct end product, the other processes yield a very low amount of nano-biochar, which is one of the key constraints in monetizing the product. Additionally, transitioning from lab-scale to industrial-scale may pose a variety of difficulties in the milling techniques. Therefore, commercializing nano-biochar requires further research and a techno-economic study (Lam et al., 2019).

3.4.1 THE OBSTACLES INVOLVED WITH THE USE OF NANO-BIOCHAR

There are several constraints to using natural nano-biochar for direct environmental applications, including low stability as well as difficulties isolating nanoparticles, easy mobility, absorption, and accumulation, and the nature of toxicity with regards to very thin particles (nano to micron) (Wang et al., 2013; Liu et al., 2018). The dispersion of nano-biochar size might be affected by a variety of parameters, including the size of the source material, mechanical grinding/milling, and aging method (Masek et al., 2018).

3.5 NANOBIOCHAR IN SOIL-PLANT IMPROVEMENTS

Aside from its use as a medium for adsorptive removal, optoelectronic uses, and energy storage, nano-biochar has been investigated for various purposes, such as solid fuel briquettes for pellet biofuels and rice bio-fertilizers (Li et al., 2017b). Water-soluble carbon-based nanoparticles derived from acid-modified nutrient-loaded (e.g., NO_3^- and NH_4^+) rice husk biochar have been shown to accelerate wheat seed germination (Saxena et al., 2014). Over 5 days, shoot growth rates increased with water-soluble carbon nanoparticle concentrations ranging from 10 to 50 mg/L. Furthermore, water-soluble carbon nanoparticle concentrations greater than 50 mg/L cause a modest reduction in growth rate when compared to a dosage of 50 mg/L. As a result, large concentrations of water-soluble carbon NPs may cause excessive stress on the pace of the development of shoots (Saxena et al., 2014). Yue et al. (2019) investigated the impact of nano-biochar on rice development and Cd^{2+} uptake. When compared to low-temperature (300 and 400°C) nano-biochar as well as macro biochar, high-pyrolysis temperature nano-biochar (500 and 600°C) resulted in a significant decrease in Cd^{2+} absorption and phytotoxicity. High-temperature nano-biochars also showed a notable reduction in Cd^{2+} toxicity and stress on rice seedlings, and induced stress on rice plants (Yue et al., 2019). Nano-biochar has been used to protect native ecosystems and plants by reducing the stress of invasive plants. Additionally, nano-biochar has shown potential in improving rice seedlings and biomass, as well as increasing root length and chlorophyll levels while reducing oxidative stress. Furthermore, the potential risks of using nano-biochar should be studied in the future (Shen et al., 2020). In this context, nano-biochar may be used as a nano-carrier for insecticides, nutrients, and plant growth chemicals to enhance production (Figure 3.3).

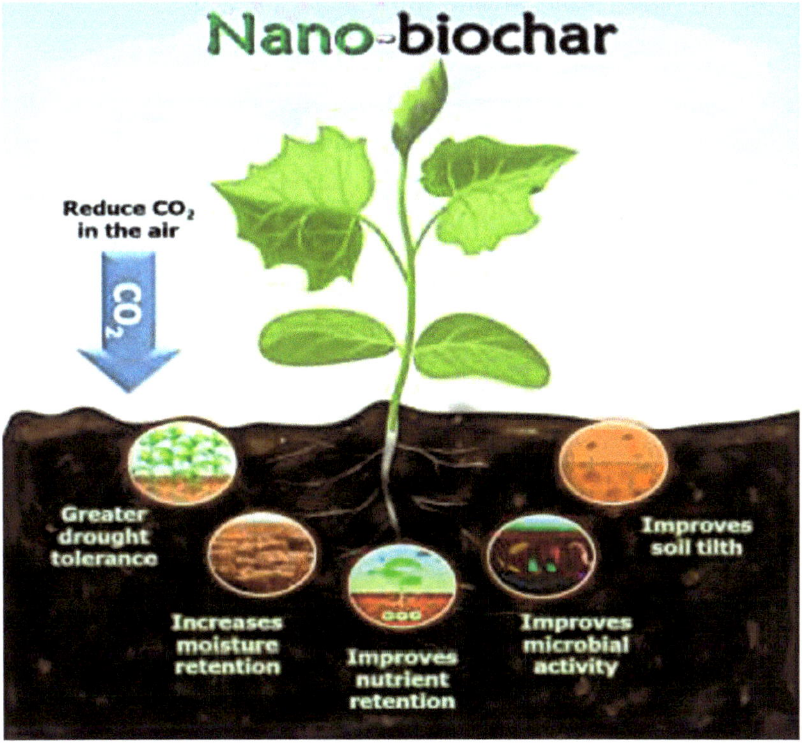

FIGURE 3.3 Benefits of applying nano-biochar to soil.

3.5.1 NANO-BIOCHAR TECHNOLOGY FOR SOIL REMEDIATION AND PHYTOTOXICITY CONTROL

Biochar has demonstrated capable potential in lowering the bioavailability of soil pollutants, including potentially hazardous metals (Hong et al., 2019; Luo et al., 2020), improving soil health (Zhang et al., 2019b), as well as increasing nutrient bioavailability (Li et al., 2020). Furthermore, only limited research has documented the application of nano-biochar to reduce metal-induced phytotoxicity and soil pollution. So far, nano-remediation techniques have contributed a substantial amount to environmental preservation and the reduction of contamination (Gil-Daz et al., 2017). Biochar and nano-biochar can manage the considerable influence of carbon nanotechnology in agriculture and plant biotechnology to increase crop output (Lian & Xing, 2017; Zhang et al., 2019a). Nano-biochar not only improves the decomposition of organic contaminants by directing electron shifts like a catalyst, but it may also directly interact with harmful pollutants, thus exerting a significant influence on environmental elements or challenges; while little information is currently available (Yang et al., 2017).

3.5.2 RISK AND TOXICITY OF NANO-BIOCHAR

The extraordinary growth of nano-biochar in recent years has drawn research interest because of its potential for expansion in a variety of applications. As a result, the harmful impacts of nano-biochar on plants and people should be extensively researched prior to large-scale agricultural or industrial uses being undertaken. These impacts can vary based on whether the biochar substance is physical or chemical (Bhati et al., 2018). Negative impacts on plant systems might be classified as cytotoxicity, genotoxicity, or ecotoxicity (Cifuentes et al., 2010). However, the toxicology of nano-biochar has yet to be properly investigated. To realize the harmful impacts on plant and human cells, detailed analyses of chemical processes and pathways are required. The effects of physico-chemical qualities on toxicity should also be investigated. Nano-biochar can sorb both organic and inorganic pollutants and migrate with surface water and groundwater, making it a potential contamination vector (Ma et al., 2019). Because nano-biochars have a wide variety of possible uses, research was done to investigate their toxicity on Streptomyces, a soil microbe which generates two-thirds of naturally occurring antibiotics. In a recent investigation, it was found that nano-biochar is more hazardous than carbon nanotubes and graphene (Liu et al., 2019). The study also revealed that the spherical form of nano-biochars, which is associated with cell damage and high ROS, resulted in the greatest antibiotic production when compared to various carbon nanomaterials. This suggests that nano-biochar may have promising future applications (Dubey et al., 2015). While interactions between plant and nano-biochar are beneficial to monocot plants, dicot plants demonstrate both favorable and unfavorable effects (Sonkar et al., 2012). The most widely held belief regarding underivatized nano-biochar toxicity is that it harms cell soft membranes. Nano-biochars, which may be disseminated in water, can be easily transferred by plant absorption via the xylem vessels (Bhati et al., 2018).

3.6 TECHNOECONOMIC EVALUATION

The initial investment, operational cost, and expected cost of preparing nano-biochar have been analyzed using prior studies on pyrolysis of biomass waste (e.g., maize cob), especially hemicellulose-rich, and employing pre-treatment procedures to improve brittleness (Lam et al., 2019). Before scaling up production, cost estimations will provide a chance to select the least expensive production technique for a certain kind of nano-biochar. Furthermore, the qualities of the desired final product must be considered since they may influence the production process used. A survey methodology or manufacturer questioning can be used to determine the cost of nano-biochar production equipment. The entire plant cost covers the total cost of core equipment, design, plant setup, manufacture, and system change if necessary. The total plant cost includes both fixed capital investment (FCI) and start-up and operating capital costs (15% of FCI) (Islam & Ani, 2000). As a result, the overall cost of manufacturing for each step is determined by the nation and the output.

3.6.1 NANO-BIOCHAR FOR SUSTAINABLE AGRICULTURE

Climate change and technological challenges, which include imbalanced resource usage and environmental difficulties such as eutrophication, surface runoff, the application of chemical fertilizer, and industrial waste emissions, have recently had a significant impact on the agricultural sector (Maria et al., 2013; Withers et al., 2014). These problems, particularly in agriculture and soil, might be remedied by employing nano-biochar (Lu et al., 2020), which can reduce the bioavailability of environmental toxins and recover damaged soils (Oni et al., 2019).

Nano-biochar, which is now used as a soil supplement, has a significant influence on soil health by modifying the physico-chemical and biological properties of the soil (Awad et al., 2018; Xiao et al., 2018). The use of nano-biochar increases soil fertility, making it more suitable for the development of plants. Nano-biochar provides an optimal soil habitat for bacteria by improving soil porosity, structural rigidity, and water retention capabilities, all of which are essential for soil activity maintenance (Ameloot et al., 2013). According to reports, nano-biochar significantly improves soil fertility by increasing microbial diversity (Liu et al., 2020b).

3.7 CONCLUSION

Nano-biochar has shown impressive adsorption capacities for both harmful metals and organic pollutants, as well as beneficial catalytic and electro-chemical properties. Moreover, nano-biochar has developed as an effective material for energy storage and conversion applications. The manufacturing of nano-biochar is an efficient and environmentally friendly technique which can improve the effectiveness of water and fertilizer usage in the soil. It may also be utilized as a medium for macro and micronutrients, allowing the nutrients to absorb all of the increased fertilizer. However, both biochar and nano-biochars can be employed to boost crop and plant health by supplying them with a high nutritional content over an extended period. The application of nano-biochar to the soil improves the resistance of organic carbon in the soil against environmental variables while decreasing mineralization. Because of their extremely tiny size and enhanced surface area, nano-biochar has the potential for environmental remediation of inorganic and organic pollutants, which enhances soil fertility and hence crop performance. As a result, biochar applications have the opportunity to play a unique and critical role in the future of sustainable agriculture. This study revealed research gaps and problems in nano-biochar research and applications in the future.

3.8 VISIONS FOR THE FUTURE

Further research is required to evaluate the %C, %H, %O, and other general characteristics like EC, zeta potential, and pH, to acquire a greater knowledge of the properties of nano-biochar and how these change with feedstock, pyrolysis temperatures, and nano-biochar manufacturing technique. In the same way, the physico-chemical and biological properties of biochar should be thoroughly examined to uncover links among nano-biochar qualities and nature. There haven't been any studies done on the toxicity of nano-biochar on human cell systems, and there haven't been many inquiries into the negative consequences of nano-biochar. Furthermore, nano-biochar has the potential to be used as a tool for the delayed release of nano fertilizers, plant nano-bionics, and medical purposes.

KEYWORDS

- **biochar**
- **carbon material**
- **nano-biochar**
- **soil amendment**

REFERENCES

Ahmad, M., Rajapaksha, A. U., Lim, J. E., Zhang, M., Bolan, N., Mohan, D., Vithanage, M., et al., (2014). Biochar as a sorbent for contaminant management in soil and water: A review. *Chemosphere, 99*, 19–33.

Alam, M. S., Gorman-Lewis, D., Chen, N., Flynn, S. L., Ok, Y. S., Konhauser, K. O., & Alessi, D. S., (2018a). Thermodynamic analysis of nickel (II) and zinc (II) adsorption to biochar. *Environmental Science & Technology, 52*(11), 6246–6255.

Alam, M. S., Gorman-Lewis, D., Chen, N., Safari, S., Baek, K., Konhauser, K. O., & Alessi, D. S., (2018b). Mechanisms of the removal of U (VI) from aqueous solution using biochar: A combined spectroscopic and modeling approach. *Environmental Science & Technology, 52*(22), 13057–13067.

Alameri, R. A., (2021). Green medicine: A novel preparation method for green synthesizing of iron nanoparticles derived from beta vulgaris extract. *Archives of Razi Institute, 76*(5), 1327.

Albiter, M. A., Crooks, R. M., & Zaera, F., (2010). Adsorption of carbon monoxide on dendrimer-encapsulated platinum nanoparticles: Liquid versus gas phase. *The Journal of Physical Chemistry Letters, 1*(1), 38–40.

Ameloot, N., Graber, E. R., Verheijen, F. G., & De Neve, S., (2013). Interactions between biochar stability and soil organisms: Review and research needs. *European Journal of Soil Science, 64*(4), 379–390.

An, Y., Lu, J., Niu, R., Li, M., Zhao, X., Huang, X., Huang, H., Garg, A., & Zhussupbekov, A., (2021). Exploring effects of novel chemical modification of biochar on soil water retention and crack suppression: Towards commercialization of production of biochar for soil remediation. *Biomass Conversion and Biorefinery, 19*, 1–4.

ArameshBoroujeni, Z., & AsadiAghbalaghi, Z., (2021). Electrochemical determination venlafaxine at NiO/GR nanocomposite modified carbon paste electrode. *Iranian Journal of Chemistry and Chemical Engineering (IJCCE), 40*(4), 1042–1053.

Ardebili, S. M., Taghipoor, A., Solmaz, H., & Mostafaei, M., (2020). The effect of nano-biochar on the performance and emissions of a diesel engine fueled with fusel oil-diesel fuel. *Fuel, 268*, 117356.

Arif, M., Ilyas, M., Riaz, M., Ali, K., Shah, K., Haq, I. U., & Fahad, S., (2017). Biochar improves phosphorus use efficiency of organic-inorganic fertilizers, maize-wheat productivity and soil quality in a low fertility alkaline soil. *Field Crops Research, 214*, 25–37.

Atkinson, C. J., Fitzgerald, J. D., & Hipps, N. A., (2010). Potential mechanisms for achieving agricultural benefits from biochar application to temperate soils: A review. *Plant and Soil, 337*(1), 1–8.

Awad, Y. M., Lee, S. S., Kim, K. H., Ok, Y. S., & Kuzyakov, Y., (2018). Carbon and nitrogen mineralization and enzyme activities in soil aggregate-size classes: Effects of biochar, oyster shells, and polymers. *Chemosphere, 198*, 40–48.

Azargohar, R., & Dalai, A. K., (2008). Steam and KOH activation of biochar: Experimental and modeling studies. *Microporous and Mesoporous Materials, 110*(2, 3), 413–421.

Banek, N. A., Abele, D. T., McKenzie, Jr. K. R., & Wagner, M. J., (2018). Sustainable conversion of lignocellulose to high-purity, highly crystalline flake potato graphite. *ACS Sustainable Chemistry & Engineering, 6*(10), 13199–13207.

Bhati, A., Tripathi, K. M., Singh, A., Sarkar, S., & Sonkar, S. K., (2018). Exploration of nano carbons in relevance to plant systems. *New Journal of Chemistry, 42*(20), 16411–16427.

Blanco-Canqui, H., (2021). Does biochar improve all soil ecosystem services? *GCB Bioenergy, 13*(2), 291–304.

Brewer, C. E., Chuang, V. J., Masiello, C. A., Gonnermann, H., Gao, X., Dugan, B., Driver, L. E., et al., (2014). New approaches to measuring biochar density and porosity. *Biomass and Bioenergy, 66*, 176–185.

Cao, X., & Harris, W., (2010). Properties of dairy-manure-derived biochar pertinent to its potential use in remediation. *Bioresource Technology, 101*(14), 5222–5228.

Charkhi, A., Kazemian, H., & Kazemeini, M., (2010). Optimized experimental design for natural clinoptilolite zeolite ball milling to produce nano powders. *Powder Technology, 203*(2), 389–396.

Chen, B., Chen, Z., & Lv, S., (2011). A novel magnetic biochar efficiently sorbs organic pollutants and phosphate. *Bioresource Technology, 102*(2), 716–723.

Cifuentes, Z., Custardoy, L., De La Fuente, J. M., Marquina, C., Ibarra, M. R., Rubiales, D., & Pérez-de-Luque, A., (2010). Absorption and translocation to the aerial part of magnetic carbon-coated nanoparticles through the root of different crop plants. *Journal of Nanobiotechnology, 8*(1), 1–8.

Das, O., & Sarmah, A. K., (2015). The love–hate relationship of pyrolysis biochar and water: A perspective. *Science of the Total Environment, 512*, 682–685.

Dempster, D. N., Jones, D. L., & Murphy, D. V., (2012). Clay and biochar amendments decreased inorganic but not dissolved organic nitrogen leaching in soil. *Soil Research, 50*(3), 216–221.

Dong, X., Guan, T., Li, G., Lin, Q., & Zhao, X., (2016). Long-term effects of biochar amount on the content and composition of organic matter in soil aggregates under field conditions. *Journal of Soils and Sediments, 16*(5), 1481–1497.

Duan, M., Liu, G., Zhou, B., Chen, X., Wang, Q., Zhu, H., & Li, Z., (2021). Effects of modified biochar on water and salt distribution and water-stable macro-aggregates in saline-alkaline soil. *Journal of Soils and Sediments, 21*(6), 2192–2202.

Dubey, P., Tripathi, K. M., Mishra, R., Bhati, A., Singh, A., & Sonkar, S. K., (2015). A simple one-step hydrothermal route towards water solubilization of carbon quantum dots from soya-nuggets for imaging applications. *RSC Advances, 5*(106), 87528–87534.

Feng, W., Yang, F., Cen, R., Liu, J., Qu, Z., Miao, Q., & Chen, H., (2021). Effects of straw biochar application on soil temperature, available nitrogen and growth of corn. *Journal of Environmental Management, 277*, 111331.

Gámiz, B., Cox, L., Hermosín, M. C., Spokas, K., & Celis, R., (2017). Assessing the effect of organoclays and biochar on the fate of abscisic acid in soil. *Journal of Agricultural and Food Chemistry, 65*(1), 29–38.

Gao, X., & Wu, H., (2014). Aerodynamic properties of biochar particles: Effect of grinding and implications. *Environmental Science & Technology Letters, 1*(1), 60–64.

Gil-Díaz, M., Pinilla, P., Alonso, J., & Lobo, M. C., (2017). Viability of a nanoremediation process in single or multi-metal (loid) contaminated soils. *Journal of Hazardous Materials, 321*, 812–819.

Guo, F., Bao, L., Wang, H., Larson, S. L., Ballard, J. H., A., (2020). Simple method for the synthesis of biochar nanodots using hydrothermal reactor. *MethodsX*. 10.1016/j.mex.2020.101022.

Han, F., Zheng, J., Hu, W., Du, F., & Zhang, X., (2010). Spatial variability and distribution of soil nutrients in a catchment of the loess plateau in China. *Acta Agriculturae Scandinavica Section B–Soil and Plant Science, 60*(1), 48–56.

He, Y., Yao, Y., Ji, Y., Deng, J., Zhou, G., Liu, R., Shao, J., et al., (2020). Biochar amendment boosts photosynthesis and biomass in C_3 but not C_4 plants: A global synthesis. *GCB Bioenergy, 12*(8), 605–617.

Honarvar, Z., Farhoodi, M., Khani, M. R., Mohammadi, A., Shokri, B., Jannatiha, N., & Shojaee-Aliabadi, S., (2021). Antimicrobial and physicochemical properties of plasma treated bio-coating polypropylene films containing *satureja hortensis* essential oil. *Iran. J. Chem. Chem. Eng. Research Article, 40*(4).

Hong, M., Zhang, L., Tan, Z., & Huang, Q., (2019). Effect mechanism of biochar's zeta potential on farmland soil's cadmium immobilization. *Environmental Science and Pollution Research, 26*(19), 19738–19748.

Huang, J., Zimmerman, A. R., Chen, H., & Gao, B., (2020). Ball milled biochar effectively removes sulfamethoxazole and sulfapyridine antibiotics from water and wastewater. *Environmental Pollution, 258*, 113809.

Islam, M. N., & Ani, F. N., (2000). Techno-economics of rice husk pyrolysis, conversion with catalytic treatment to produce liquid fuel. *Bioresource Technology, 73*(1), 67–75.

Jiang, C., Bo, J., Xiao, X., Zhang, S., Wang, Z., Yan, G., Wu, Y., et al., (2020). Converting waste lignin into nano-biochar as a renewable substitute of carbon black for reinforcing styrene-butadiene rubber. *Waste Management, 102*, 732–742.

Jun, F., Yu, G., Quanjiu, W., Malhi, S. S., & Yangyang, L., (2014). Mulching effects on water storage in soil and its depletion by alfalfa in the Loess Plateau of Northwestern China. *Agricultural Water Management, 138*, 10–16.

Khorassani, R., Hettwer, U., Ratzinger, A., Steingrobe, B., Karlovsky, P., & Claassen, N., (2011). Citramalic acid and salicylic acid in sugar beet root exudates solubilize soil phosphorus. *BMC Plant Biology, 11*(1), 1–8.

Kianfar, E., (2021). Magnetic nanoparticles in targeted drug delivery: A review. *Journal of Superconductivity and Novel Magnetism, 34*(7), 1709–1735.

Laird, D. A., Fleming, P., Davis, D. D., Horton, R., Wang, B., & Karlen, D. L., (2010). Impact of biochar amendments on the quality of a typical Midwestern agricultural soil. *Geoderma, 158*(3, 4), 443–449.

Lam, S. S., Mahari, W. A., Ok, Y. S., Peng, W., Chong, C. T., Ma, N. L., Chase, H. A., et al., (2019). Microwave vacuum pyrolysis of waste plastic and used cooking oil for simultaneous waste reduction and sustainable energy conversion: Recovery of cleaner liquid fuel and techno-economic analysis. *Renewable and Sustainable Energy Reviews, 115*, 109359.

Li, H., Qiu, Y. F., Wang, X. L., Yang, J., Yu, Y. J., & Chen, Y. Q., (2017b). Biochar supported Ni/Fe bimetallic nanoparticles to remove 1, 1, 1-trichloroethane under various reaction conditions. *Chemosphere, 169*, 534–541.

Li, R., Zhang, Y., Deng, H., Zhang, Z., Wang, J. J., Shaheen, S. M., Xiao, R., et al., (2020). Removing tetracycline and Hg (II) with ball-milled magnetic nanobiochar and its potential on polluted irrigation water reclamation. *Journal of Hazardous Materials, 384*, 121095.

Li, X., Liu, L., Wang, X., Ok, Y. S., Elliott, J. A., Chang, S. X., & Chung, H. J., (2017a). Flexible and self-healing aqueous supercapacitors for low temperature applications: Polyampholyte gel electrolytes with biochar electrodes. *Scientific Reports, 7*(1), 1–1.

Lian, F., & Xing, B., (2017). Black carbon (biochar) in water/soil environments: Molecular structure, sorption, stability, and potential risk. *Environmental Science & Technology, 51*(23), 13517–13532.

Lian, F., Yu, W., Wang, Z., & Xing, B., (2018). New insights into black carbon nanoparticle-induced dispersibility of goethite colloids and configuration-dependent sorption for phenanthrene. *Environmental Science & Technology, 53*(2), 661–670.

Liu, B., Li, H., Li, H., Zhang, A., & Rengel, Z., (2021a). Long-term biochar application promotes rice productivity by regulating root dynamic development and reducing nitrogen leaching. *GCB Bioenergy, 13*(1), 257–268.

Liu, C., Sun, B., Zhang, X., Liu, X., Drosos, M., Li, L., & Pan, G., (2021b). The water-soluble pool in biochar dominates maize plant growth promotion under biochar amendment. *Journal of Plant Growth Regulation, 40*(4), 1466–1476.

Liu, G., Zheng, H., Jiang, Z., Zhao, J., Wang, Z., Pan, B., & Xing, B., (2018). Formation and physicochemical characteristics of nano biochar: Insight into chemical and colloidal stability. *Environmental Science & Technology, 52*(18), 10369–10379.

Liu, J., Schulz, H., Brandl, S., Miehtke, H., Huwe, B., & Glaser, B., (2012). Short-term effect of biochar and compost on soil fertility and water status of a dystric cambisol in NE Germany under field conditions. *Journal of Plant Nutrition and Soil Science, 175*(5), 698–707.

Liu, W., Li, Y., Feng, Y., Qiao, J., Zhao, H., Xie, J., Fang, Y., et al., (2020a). The effectiveness of nanobiochar for reducing phytotoxicity and improving soil remediation in cadmium-contaminated soil. *Scientific Reports, 10*(1), 1–0.

Liu, X., Tang, J., Wang, L., Liu, Q., & Liu, R., (2019). A comparative analysis of ball-milled biochar, graphene oxide, and multi-walled carbon nanotubes with respect to toxicity induction in *Streptomyces*. *Journal of Environmental Management, 243*, 308–317.

Liu, Z., Wu, X., Liu, W., Bian, R., Ge, T., Zhang, W., Zheng, J., et al., (2020b). Greater microbial carbon use efficiency and carbon sequestration in soils: Amendment of biochar versus crop straws. *GCB Bioenergy, 12*(12), 1092–1103.

Lu, K., Yang, X., Gielen, G., Bolan, N., Ok, Y. S., Niazi, N. K., Xu, S., et al., (2017). Effect of bamboo and rice straw biochars on the mobility and redistribution of heavy metals (Cd, Cu, Pb and Zn) in contaminated soil. *Journal of Environmental Management, 186*, 285–292.

Lu, L., Yu, W., Wang, Y., Zhang, K., Zhu, X., Zhang, Y., Wu, Y., et al., (2020). Application of biochar-based materials in environmental remediation: From multi-level structures to specific devices. *Biochar, 2*(1), 1–31.

Luo, M., Lin, H., He, Y., & Zhang, Y., (2020). The influence of corncob-based biochar on remediation of arsenic and cadmium in yellow soil and cinnamon soil. *Science of the Total Environment, 717*, 137014.

Luo, Y., Lin, Q., Durenkamp, M., Dungait, A. J., & Brookes, P. C., (2017a). Soil priming effects following substrates addition to biochar-treated soils after 431 days of pre-incubation. *Biology and Fertility of Soils, 53*(3), 315–326.

Luo, Y., Zang, H., Yu, Z., Chen, Z., Gunina, A., Kuzyakov, Y., Xu, J., et al., (2017b). Priming effects in biochar enriched soils using a three-source-partitioning approach: 14C labelling and 13C natural abundance. Soil *Biology and Biochemistry, 106*, 28–35.

Lyu, H., Gao, B., He, F., Zimmerman, A. R., Ding, C., Huang, H., & Tang, J., (2018). Effects of ball milling on the physicochemical and sorptive properties of biochar: Experimental observations and governing mechanisms. *Environmental Pollution, 233*, 54–63.

Ma, S., Jing, F., Sohi, S. P., & Chen, J., (2019). New insights into contrasting mechanisms for PAE adsorption on millimeter, micron-and nano-scale biochar. *Environmental Science and Pollution Research, 26*(18), 18636–18650.

Mahmoud, E., El Baroudy, A., Ali, N., & Sleem, M., (2020). Spectroscopic studies on the phosphorus adsorption in salt-affected soils with or without nano-biochar additions. *Environmental Research, 184*, 109277.

Majdi, H. S., Latipov, Z. A., Borisov, V., Yuryevna, N. O., Kadhim, M. M., Suksatan, W., Khlewee, I. H., & Kianfar, E., (2021). Nano and battery anode: A review. *Nanoscale Research Letters, 16*(1), 1–26.

Manyà, J. J., (2012). Pyrolysis for biochar purposes: A review to establish current knowledge gaps and research needs. *Environmental Science & Technology, 46*(15), 7939–7954.

Maria, D. S., Diego, B., & Giovanni, S., (2013). Measuring the effect of climate change on agriculture: A literature review of analytical models. *Journal of Development and Agricultural Economics, 5*(12), 499–509.

Masek, O., Buss, W., & Sohi, S., (2018). Standard biochar materials. *Environmental Science & Technology, 52*(17), 9543–9544.

Meng, L., Sun, T., Li, M., Saleem, M., Zhang, Q., & Wang, C., (2019). Soil-applied biochar increases microbial diversity and wheat plant performance under herbicide fomesafen stress. *Ecotoxicology and Environmental Safety, 171*, 75–83.

Meyer, S., Glaser, B., & Quicker, P., (2011). Technical, economical, and climate–related aspects of biochar production technologies: A literature review. *Environ. Sci. Technol., 45*, 9473–9483.

Mohan, D., Kumar, S., & Srivastava, A., (2014). Fluoride removal from ground water using magnetic and nonmagnetic corn stover biochars. *Ecol. Eng., 73*, 798–808.

Moradi, N., & Karimi, A., (2021). Fe-Modified common reed biochar reduced cadmium (Cd) mobility and enhanced microbial activity in a contaminated calcareous soil. *Journal of Soil Science and Plant Nutrition, 21*(1), 329–340.

Naghdi, M., Taheran, M., Brar, S. K., Kermanshahi-Pour, A., Verma, M., & Surampalli, R. Y., (2017a). Immobilized laccase on oxygen functionalized nanobiochars through mineral acids treatment for removal of carbamazepine. *Science of The Total Environment, 584*, 393–401.

Naghdi, M., Taheran, M., Brar, S. K., Kermanshahi-Pour, A., Verma, M., & Surampalli, R. Y., (2018). Pinewood nanobiochar: A unique carrier for the immobilization of crude laccase by covalent bonding. *International Journal of Biological Macromolecules, 115*, 563–571.

Naghdi, M., Taheran, M., Brar, S. K., Rouissi, T., Verma, M., Surampalli, R. Y., & Valero, J. R., (2017b). A green method for production of nanobiochar by ball milling-optimization and characterization. *Journal of Cleaner Production, 164*, 1394–1405.

Naghdi, M., Taheran, M., Pulicharla, R., Rouissi, T., Brar, S. K., Verma, M., & Surampalli, R. Y., (2019). Pine-wood derived nanobiochar for removal of carbamazepine from aqueous media: Adsorption behavior and influential parameters. *Arabian Journal of Chemistry, 12*(8), 5292–5301.

Nath, B. K., Chaliha, C., & Kalita, E., (2019). Iron oxide permeated mesoporous rice-husk nanobiochar (IPMN) mediated removal of dissolved arsenic (As): Chemometric modelling and adsorption dynamics. *Journal of Environmental Management, 246*, 397–409.

Oleszczuk, P., Ćwikła-Bundyra, W., Bogusz, A., & Skwarek, E., (2016). Characterization of nanoparticles of biochars from different biomass. *J. Anal. Appl. Pyrolysis, 121*, 165–117.

Oleszczuk, P., Ćwikła-Bundyra, W., Bogusz, A., Skwarek, E., & Ok, Y. S., (2016). Characterization of nanoparticles of biochars from different biomass. *Journal of Analytical and Applied Pyrolysis, 121*, 165–172.

Oni, B. A., Oziegbe, O., & Olawole, O. O., (2019). Significance of biochar application to the environment and economy. *Annals of Agricultural Sciences, 64*(2), 222–236.

Qian, L., Zhang, W., Yan, J., Han, L., Gao, W., Liu, R., & Chen, M., (2016). Effective removal of heavy metal by biochar colloids under different pyrolysis temperatures. *Bioresource Technology, 206*, 217–224.

Qu, X., Fu, H., Mao, J., Ran, Y., Zhang, D., & Zhu, D., (2016). Chemical and structural properties of dissolved black carbon released from biochars. *Carbon, 96*, 759–767.

Ramanayaka, S., Vithanage, M., Alessi, D., Liu, W. J., & Jayasundera, A. C. A., (2020). Nanobiochar: Production, properties, and multifunctional applications. *Environ. Sci. Nano, 7*, 3279–3302.

Rens, H., Bera, T., & Alva, A. K., (2018). Effects of biochar and biosolid on adsorption of nitrogen, phosphorus, and potassium in two soils. *Water, Air, & Soil Pollution, 229*(8), 1–3.

Rhoades, C. C., Minatre, K. L., Pierson, D. N., Fegel, T. S., Cotrufo, M. F., & Kelly, E. F., (2017). Examining the potential of forest residue-based amendments for post-wildfire rehabilitation in Colorado, USA. *Scientifica.*

Safari, S., Von, G. K., Alam, M., Hubmann, M., Blewett, T. A., Chi, Z., & Alessi, D. S., (2019). Biochar colloids and their use in contaminants removal. *Biochar, 1*(2), 151–162.

Saffari, M., Karimian, N., Ronaghi, A., Yasrebi, J., & Ghasemi, F. R., (2015). *Immobilization of Cadmium in a Cd-Spiked Soil by Different Kinds of Amendments,* 221–233.

Saied, S. M., Mohammed, S. J., Khaleel, B. T., & Saleh, M. Y., (2021). Comparative studies between conventional techniques and green chemistry to synthesis of novel piperidinium salts ionic liquids (PBSILs). *Journal of Chemical Health Risks, 11*(4), 451–456.

Salim, F. B., Shiaka, G. P., Muhammad, M., Nafisa, B., Surayya, M. M., Sa'adatu, A. Y., Gayus, R., & Gumel, A. M., (2021). Influence of reaction temperature on bioethanol production by *Saccharomyces cerevisiae* using cassava as substrate. *International Journal of Sustainable Energy and Environmental Research, 10*(1), 9–16.

Saxena, M., Maity, S., & Sarkar, S., (2014). Carbon nanoparticles in 'biochar' boost wheat (*Triticum aestivum*) plant growth. *RSC Advances, 4*(75), 39948–39954.

Shah, U. V., Wang, Z., Olusanmi, D., Narang, A. S., Hussain, M. A., Tobyn, M. J., & Heng, J. Y., (2015). Effect of milling temperatures on surface area, surface energy and cohesion of pharmaceutical powders. *International Journal of Pharmaceutics, 495*(1), 234–240.

Sharma, R., Bell, R. W., & Wong, M. T., (2015). Phosphorus forms in soil solution and leachate of contrasting soil profiles and their implications for P mobility. *Journal of Soils and Sediments, 15*(4), 854–862.

Shen, Y., Tang, H., Wu, W., Shang, H., Zhang, D., Zhan, X., & Xing, B., (2020). Role of nano-biochar in attenuating the allelopathic effect from *Imperata cylindrica* on rice seedlings. *Environmental Science: Nano, 7*(1), 116–126.

Sohi, S. P., Krull, E., Lopez-Capel, E., & Bol, R., (2010). A review of biochar and its use and function in soil. *Advances in Agronomy, 105*, 47–82.

Song, B., Chen, M., Zhao, L., Qiu, H., & Cao, X., (2019). Physicochemical property and colloidal stability of micron-and nano-particle biochar derived from a variety of feedstock sources. *Science of the Total Environment, 661*, 685–695.

Sonkar, S. K., Roy, M., Babar, D. G., & Sarkar, S., (2012). Water soluble carbon nano-onions from wood wool as growth promoters for gram plants. *Nanoscale, 4*(24), 7670–7675.

Spokas, K. A., Novak, J. M., Masiello, C. A., Johnson, M. G., Colosky, E. C., Ippolito, J. A., & Trigo, C., (2014). Physical disintegration of biochar: An overlooked process. *Environmental Science & Technology Letters, 1*(8), 326–332.

Syah, R., Zahar, M., & Kianfar, E., (2021). Nanoreactors: Properties, applications and characterization. *International Journal of Chemical Reactor Engineering, 19*(10), 981–1007.

Vithanage, M., Rajapaksha, A. U., Zhang, M., Thiele-Bruhn, S., Lee, S. S., & Ok, Y. S., (2015). Acid-activated biochar increased sulfamethazine retention in soils. *Environmental Science and Pollution Research, 22*(3), 2175–2186.

Wallace, C. A., Afzal, M. T., & Saha, G. C., (2019). Effect of feedstock and microwave pyrolysis temperature on physio-chemical and nano-scale mechanical properties of biochar. *Bioresources and Bioprocessing, 6*(1), 1–1.

Wang, B., Gao, B., & Fang, J., (2017a). Recent advances in engineered biochar productions and applications. *Critical Reviews in Environmental Science and Technology, 47*(22), 2158–2207.

Wang, D., Zhang, W., & Zhou, D., (2013). Antagonistic effects of humic acid and iron oxyhydroxide grain-coating on biochar nanoparticle transport in saturated sand. *Environmental Science & Technology, 47*(10), 5154–5161.

Wang, Y., Wei, Y., & Sun, J., (2016). Biochar application promotes growth parameters of soybean and reduces the growth difference. *Communications in Soil Science and Plant Analysis, 47*(12), 1493–1502.

Wang, Z. Y., Chen, L., Sun, F. L., Luo, X. X., Wang, H. F., Liu, G. C., Xu, Z. H., et al., (2017b). Effects of adding biochar on the properties and nitrogen bioavailability of an acidic soil. *European Journal of Soil Science, 68*(4), 559–572.

Weber, K., & Quicker, P., (2018). Properties of biochar. *Fuel, 217*, 240–261.

Withers, P. J., Neal, C., Jarvie, H. P., & Doody, D. G., (2014). Agriculture and eutrophication: Where do we go from here? *Sustainability, 6*(9), 5853–5875.

Woods, W. I., Teixeira, W. G., Lehmann, J., Steiner, C., WinklerPrins, A., & Rebellato, L., (2009). *Amazonian Dark Earths: Wim Sombroek's Vision.* Berlin: Springer.

Wrobel-Tobiszewska, A., Boersma, M., Adams, P., Singh, B., Franks, S., & Sargison, J., (2014). Biochar for eucalyptus forestry plantations. In: *XXIX International Horticultural Congress on Horticulture: Sustaining Lives, Livelihoods and Landscapes (IHC2014).*

Wu, L., Wei, C., Zhang, S., Wang, Y., Kuzyakov, Y., & Ding, X., (2019). MgO-modified biochar increases phosphate retention and rice yields in saline-alkaline soil. *Journal of Cleaner Production, 235*, 901–909.

Xiao, D., Yuan, D., He, H., & Gao, M., (2013). Microwave assisted one-step green synthesis of fluorescent carbon nanoparticles from ionic liquids and their application as novel fluorescence probe for quercetin determination. *Journal of Luminescence, 140*, 120–125.

Xiao, X., Chen, B., Chen, Z., Zhu, L., & Schnoor, J. L., (2018). Insight into multiple and multilevel structures of biochars and their potential environmental applications: A critical review. *Environmental Science & Technology, 52*(9), 5027–5047.

Xu, X., Zheng, Y., Gao, B., & Cao, X., (2019). N-doped biochar synthesized by a facile ball-milling method for enhanced sorption of CO_2 and reactive red. *Chemical Engineering Journal, 368*, 564–572.

Xu, Y., Seshadri, B., Sarkar, B., Wang, H., Rumpel, C., Sparks, D., Farrell, M., et al., (2018). Biochar modulates heavy metal toxicity and improves microbial carbon use efficiency in soil. *Science of the Total Environment, 621*, 148–159.

Yang, J., Pignatello, J. J., Pan, B., & Xing, B., (2017). Degradation of p-nitrophenol by lignin and cellulose chars: H_2O_2-mediated reaction and direct reaction with the char. *Environmental Science & Technology, 51*(16), 8972–8980.

Yang, T., Xu, Y., Huang, Q., Sun, Y., Liang, X., Wang, L., Qin, X., & Zhao, L., (2021). An efficient biochar synthesized by iron-zinc modified corn straw for simultaneously immobilization Cd in acidic and alkaline soils. *Environmental Pollution, 291*, 118129.

Yi, P., Pignatello, J. J., Uchimiya, M., & White, J. C., (2015). Heteroaggregation of cerium oxide nanoparticles and nanoparticles of pyrolyzed biomass. *Environmental Science & Technology, 49*(22), 13294–13303.

Yuan, Y., Bolan, N., Prévoteau, A., Vithanage, M., Biswas, J. K., Ok, Y. S., & Wang, H., (2017). Applications of biochar in redox-mediated reactions. *Bioresource Technology, 246*, 271–281.

Yue, L., Lian, F., Han, Y., Bao, Q., Wang, Z., & Xing, B., (2019). The effect of biochar nanoparticles on rice plant growth and the uptake of heavy metals: Implications for agronomic benefits and potential risk. *Science of the Total Environment, 656,* 9–18.

Yuvaraj, M., & Subramanian, K. S., (2018). Development of slow release Zn fertilizer using nano-zeolite as carrier. *Journal of Plant Nutrition, 41*(3), 311–320.

Zhang, J. H., Jian-Li, L. I., Zhang, J. B., Fu-Tao, Z. H., Cheng, Y. N., & Wei-Peng, W. A., (2010). Effects of nitrogen application rates on translocation of dry matter and nitrogen utilization in rice and wheat. *Acta Agronomica Sinica, 36*(10), 1736–1742.

Zhang, K., Mao, J., & Chen, B., (2019a). Reconsideration of heterostructures of biochars: Morphology, particle size, elemental composition, reactivity and toxicity. *Environmental Pollution, 254*, 113017.

Zhang, L., Xiang, Y., Jing, Y., & Zhang, R., (2019b). Biochar amendment effects on the activities of soil carbon, nitrogen, and phosphorus hydrolytic enzymes: A meta-analysis. *Environmental Science and Pollution Research, 26*(22), 22990–3001.

Zhang, Y., Chen, Z., Chen, C., Li, F., & Shen, K., (2021). Effects of UV-modified biochar derived from phytoremediation residue on Cd bioavailability and uptake in *Coriandrum sativum* L. in a Cd-contaminated soil. *Environmental Science and Pollution Research, 28*(14), 17395–17404.

CHAPTER 4

Role of Nano-Biochar and Biochar-Based Nanocomposites for Improving Soil Health: Application and Benefits

SHEETAL KUMARI and MANOJ CHANDRA GARG

Amity Institute of Environmental Science (AIES), Amity University, Noida, Uttar Pradesh, India

ABSTRACT

Advanced nanostructured materials, including nanobiochar, have provided long-term solutions for several contemporary challenges. Carbon nanoparticles have recently been produced as effective tools due to their unique features and a wide variety of uses in industries such as energy, agriculture, and the environment. A stable byproduct of the carbonization or pyrolysis of biomass obtained from plants and animals is biochar. Nanobiochar is a nano-sized biochar substance which has improved physical surface and chemical properties. Increasing interest in the useful uses of nanobiochar has created new opportunities for research and engineering in interdisciplinary fields. The primary determinants of its features are the feedstock type, pyrolysis temperature, residence time, and heat transfer rate. The efficiency of biochar in pollutant removal depends on its pore size, surface area, and ion exchange capacity. The key benefits of nanobiochar include carbon sequestration, improved soil fertility, pollutant removal, and recycling of agricultural byproducts and garbage. Nanobiochar is a promising alternative to biochar for waste management since it is much better at absorbing toxins, nutrients, and pollutants than biochar is, and because it is mobile in soil.

Nanomaterials and Nano-Biochar in Reducing Soil Stress: An Integrated Approach to Sustainable Agriculture. Vishnu D. Rajput, & Tatiana Minkina (Eds.)

4.1 INTRODUCTION

The term 'biochar' was coined by Read to describe the charcoal used for soil enhancement (Read et al., 2004). There has been growing interest in the potential and uses of biochar for carbon, particularly in terms of environmental management and reducing carbon dioxide emissions. One of the main issues facing the world right now is the food problem. The demand for food and agricultural goods has grown significantly due to the increase in the world population. As a result, modern agriculture is crucial for meeting the global demand for food and agricultural goods (Pang et al., 2018). The global population is expected to reach 8.6 billion in 2030 and 9.8 billion in 2050, respectively (da Costa et al., 2019). To address this global issue, the United Nations established Sustainable Development Goal 2. By 2030, its objective includes eradicating hunger, ensuring food security and improved nutrition, and promoting sustainable agriculture. With billions of dollars spent on fertilizers, attention has been focused on agriculture and agro-food sectors, primarily on improving crop yield and ensuring food security. In the last few decades, the use of nanoscale (>100 nm) in the restoration of contaminated soils has attracted significant attention due to the development of environmental molecular science and engineering. The use of nanomaterials in the binding stages has been well reported for the removal of both organic and inorganic pollutants from water and soil (Serrà et al., 2020). In addition to nanoparticles, recently developed nanocomposites are extremely affordable, plentiful, and environmentally benign due to their primary components, which have been shown to have a wide range of application prospects in the decontamination of polluted regions. The ability of the soil to sustain the growth of agricultural plants is known as soil fertility. Soil fertility is typically defined by the physical, biological, and chemical qualities of the soil (Igalavithana et al., 2015). The quality of the soil may be impacted by soil degradation, such as reduction in water retention, salinization, and nutrient depletion, which limits food production. Over 10% of the world's agricultural land has recovered from degradation, while 44% of it moderately deteriorates and 25% is categorized as extremely degraded (El-Naggar et al., 2019). One of the primary components used to increase the fertility of the soil is fertilizer because plant development and yield depend upon both micronutrients and macronutrients. Particularly, macronutrients are nitrogen (N), phosphorus (P), and potassium (K). In addition, only 50 to 60% of N, K, and 10–25% of P applied directly to crops results in absorption by the plants (Bi et al., 2020).

The basis for plant growth and development, as well as the habitat for a variety of microorganisms, makes soil an important component of the ecosystem. Biochar, as a soil amendment, alters the chemical, biological, and physical characteristics of the soil, which has a significant effect on soil fertility (Awad et al., 2018). The efficiency of biochar may be influenced by its source, manufacturing method, the type of crop being planted, the soil characteristics, and so on. Agricultural waste, animal manure, and paper-based goods are the main feedstocks used to make biochar. The significance of these wastes for producing biochar is a practical method of converting trash into a substance of value (Brewer et al., 2014). The pyrolysis system is used for both low and high-scale production. Pyrolysis is a thermochemical process that transforms biomass into biochar, bio-oil, and syngas in the absence of air at temperatures between 350 and 700C (Varma et al., 2018). The ability of biochar technology to mitigate climate change, get rid of pollutants, and improve plant growth and soil fertility is noteworthy. Although biochar-based fertilizer performs well in terms of regulated and slow release, the sustained production mechanism of the nutrient has some problems and needs more research. Encapsulating technology and biochar infused with nutrients have been suggested to enhance the sustained release mechanism (An et al., 2021). Depending on the pyrolysis technology, biochar's particle size can range from micrometers to centimeters (Lehmann, 2012). Reducing particle size in the micron range (10–600 μm) leads to more adsorption sites, which improves adsorption capacity (Naghdi et al., 2019a; Taheran et al., 2016). As a result, nano biochar is currently popular since it combines the benefits of nanotechnology with biochar. According to the literature, the pyrolysis temperature or exfoliation controls the nanoscale range's particle size (Chen et al., 2017).

On the other hand, nano biochar has been developed to mitigate biochar limitations by enhancing its structural, chemical, and physical capabilities. For instance, as observed recently by Noreen and Abd-Elsalam, it is advantageous and distinctive because of its increased surface area caused by its abundance of functional groups with greater thermal and mechanical stability, and nano size (Noreen & Abd-Elsalam, 2021). The use of biochar-based materials as adsorbents has been well-studied for several applications. In addition, they play an important role in carbon sequestration, plant development, and minimizing global warming (Naghdi et al., 2018). Nano biochar is used to create nanocomposites for wastewater treatment, adsorbing organic pollutants, and energy production (Naghdi et al., 2018). Recently, due to its porous nature, nano biochar has become a potential

negative zeta potential, and more functional groups that contain oxygen (Ramanayaka, Tsang et al., 2020; Ramanayaka, Vithanage et al., 2020). Moreover, the remediation of contaminated soils and water with the use of nano biochar has been the subject of many studies. Nano biochar can be used to remove impurities from polluted water as well as from polluted soils (Sun et al., 2022).

4.3 PREPARATION OF NANO BIOCHAR

Typically, the physiochemical features of nanobiochar will depend on its raw material. According to Weber & Quicker (2018), biomass with a high hemicellulose percentage typically produces nanobiochar with a low carbon content and a high oxygen content, whereas materials with a high lignin concentration typically form nanobiochar with a strong aggregation ability (Föhr, 2017). Researchers are interested in using biochar to improve the environment since it is produced in an economical, sustainable, and recyclable manner (Gupta et al., 2021; Mishra et al., 2020). Temperature is the primary abiotic factor that influences the synthesis of biochar (Kambo & Dutta, 2015). Standard thermochemical processes used to produce biochar include pyrolysis, gasification, hydrothermal carbonization, and torrefaction (Yuan et al., 2017). Depending on the temperature and retention time, this process can also be categorized as slow, moderate, or fast pyrolysis (Mohan et al., 2006). All thermochemical processes are shown in Table 4.1. The several pyrolysis steps used to produce biochar include endothermic pyrolysis, exothermic pyrolysis, drying and conditioning, and torrefaction. The characteristics of the biomass used and the processing factors, including temperature, heating duration, residence time, etc., affect how biochar is formed (Yaashika et al., 2020). Usually, bulk biochar is produced using a few more procedures than nanobiochar (Chausali et al., 2021).

TABLE 4.1 Thermochemical Process for Biochar with Char Yield

Process	Residence time	Temperature oC	Char yield	References
Fast Pyrolysis	less than 1s	500–1,000	15	Gurwick et al. (2013)
Slow Pyrolysis	hours to days	300–700	37	Gai et al. (2014)
Gasification	5–25 s	750–1,000	10	Leng et al. (2015)
Hydrothermal	1–12 h	180–250	30–60	Bolan et al. (2013)

4.4 AMENDED NANO BIOCHAR

Recently, several techniques have been used to modify bulk biochar and nano biochar to enhance their qualities (Wang et al., 2021). Furthermore, lignocellulosic material was shown to yield more biochar than municipal solid waste. According to studies (Ashiq et al., 2019), the performance of biochar is also improved by various physical and chemical modification processes. Chemical oxidation, CO_2 activation, steam coating, acid/base treatment, impregnation with natural and synthetic nanoparticles, etc., are examples of chemical modification processes. However, physical modification methods like ball milling have not gained favor despite the limited research that has been done in this area, which has produced positive results and calls for future study (Lyu et al., 2018; Ramanayaka, Tsang et al., 2020).

4.5 MILLING METHOD

For the preparation of nano biochar, a variety of production techniques, including ball milling, sonication, and carbonization, have been used in the literature. Moreover, several authenticated employed other techniques like centrifugation and grinding (García-Sánchez et al., 2020). The production of nanosized particles often involves mechanical grinding. Graphitic nanosheets were produced via direct nano biochar manufacture by flash heating despite mechanical procedure. To physically scatter the biochar, used an ultrasonic vibrator. Sonication was then used to create nanoparticles. The most acceptable and preferable process for producing biochar, according to the literature currently accessible, is ball milling (Oleszczuk et al., 2016). There are currently many ways to create nanobiochar, each with its own benefits and drawbacks (Liu et al., 2018). The biochar was classified into nano size using the ball milling technique by the contact of metallic balls as shown in figure 3. This technique has two milling modes: dry milling and wet milling. The difference between the two is whether the biochar is combined with water or not while being ground. Smaller and more uniformly dispersed biochar was produced by wet ball milling for 12 hours compared to dry ball milling. Both ball milling techniques also produced more biochar with a given specific surface area than manual grinding (Yuan et al., 2020). The purpose of the ball milling method is to eliminate the need for hazardous chemically assisted operations, increase atom economy, reduce waste, improve product biodegradability, and utilize renewable energy sources (Kumar et al., 2020). Ball milling is an environmentally friendly, inexpensive, repeatable procedure, but it is not the only milling technique utilized to produce nano biochar (Kumar et al., 2020). The choice of an appropriate milling technique for synthesizing nano biochar often depends on the experimental goals and the raw materials (Figure 4.3).

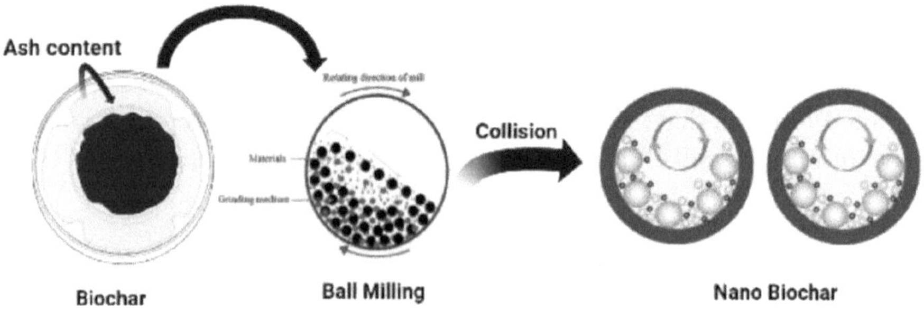

FIGURE 4.3 Preparation of nano biochar by ball milling method.

4.6 ULTRASONICATION METHOD

Ultrasonication is another successful technique for synthesizing nanobiochar (Figure 4.4). High-energy ultrasound used in the sonication process causes biochar suspended in the suspension to disintegrate into smaller pieces (Li et al., 2013). The synthesis of nanobiochar is facilitated by the exfoliation of tiny fragments that are embedded in the pores or attached to the surface of biochar. Depending on the amount of heterogeneous carbon in biochar, longer sonication times result in the production of more nanobiochar (Liu et al., 2018). Two key advantages of sonication are the uniform surface of biochar and the creation of pores without clogging (Pang et al., 2018). The main mechanism involves ultrasound-induced microjets and shockwaves that change the structure of the biochar, release closed pores, and create new pores, all of which lead to an increase in microporosity (Sajjadi et al., 2020). The properties of the biochar produced by pyrolyzing various biomass at temperatures ranging from 300–700C are variable. Unfortunately, it is difficult to make any inferences about how the sonication process parameters affect nanobiochar synthesis and characteristics because the data is not readily available.

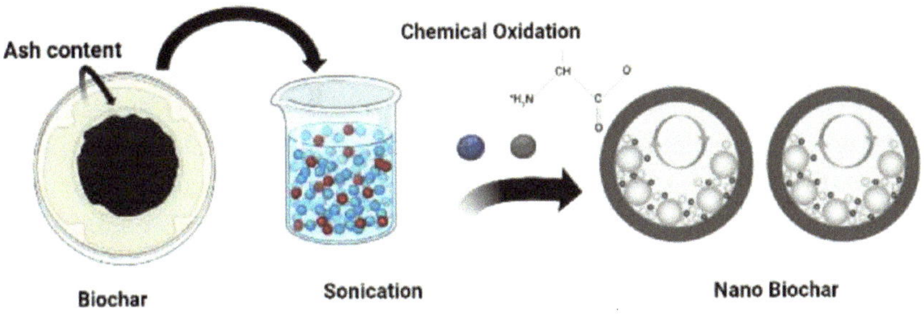

FIGURE 4.4 Production of nanobiochar by sonication method.

4.7 CARBONIZATION

The carbonization method was supplemented with a post or pre-treatment to increase the surface characteristics and size of the nanobiochar and get rid of impregnating salts to make nanobiochar. From the carbonization of waste lignin, only a few investigations have documented the formation of nanobiochar (Makshut, 2020). The lignin was devolatilized and dehydrated at high temperatures, resulting in thicker walls and a smaller cavity in the hollow structure made up of an aggregation of primary lignin particles. The primary factors in the creation of nanobiochar with a compact stack structure were the collapsing of pores and the thickening of the carbon matrix (Xu & Cölfen, 2021). Another direct approach or procedure for producing nanobiochar and the resulting ordered biochar nanoparticles is centrifugation technology. Nanobiochar can also be created via nitric and sulfuric acid decomposition of concentrated bulk biochar (Guo et al., 2020).

4.8 MECHANISM OF NANOBIOCHAR

On the basis of the multiple processes and mechanisms of lignocellulose breakdown during pyrolysis, plant-based biomass can be transformed into biochar (Yaashikaa et al., 2020). At a temperature of about 315°C, cellulose starts to degrade, and by 400°C, it has finished. Two different pathways – exothermic and endothermic – can be used for cellulose's degradation (Lehmann, 2015). Hemicellulose is the first component of plant biomass to degrade. Around 220°C, the decomposition starts, and it is finished by 315°C. Lignin degradation is complete at 160°C to 900°C. Lignin degradation is more difficult compared to cellulose and hemicellulose breakdown. Biochar, syngas, and bio-oil are formed by further processing of hemicellulose and lignin (Yaashikaa et al., 2020).

4.9 APPLICATION AND PROPERTIES OF NANO BIOCHAR

According to studies, nano biochar resembles carbon-based compounds in several ways (Chen et al., 2011). The synthesis method and feedstock materials used have a significant impact on the properties of nano biochar. FTIR research has established that lignin-derived nano biochar contains a variety of functional groups, including ether, carbonyl, hydroxyl, and aromatic characteristics. Furthermore, it has been revealed that the structure of nano biochar contains phenolic, hydroxyl stretching, and bending (Jiang et al., 2020). Hence, precise physicochemical characterization of nano biochar would aid in choosing their applications for the intended uses. Chemical addition, pyrolysis, and grinding are examples of treatment methods that can expose functional groups as well as generate new functional moieties. Pyrolysis temperature greatly influences the characteristics, morphology, and structural changes of biochar. In general, the surface area grows as the pyrolysis temperature rises. Yet, it has also been shown that the surface area at temperatures above 700°C has decreased. The aromatic lignin core may become revealed when aliphatic alkyl and ester groups are eliminated as a result of higher pyrolysis temperatures (Uchimiya et al., 2011). Applications of nanobiochar are shown in Figure 4.5.

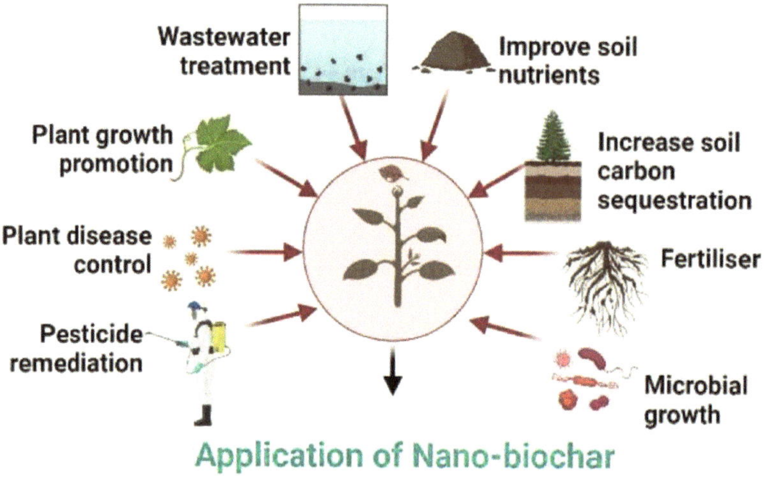

FIGURE 4.5 Different applications of nanobiochar.

4.9.1 REDUCE EMISSIONS OF GREENHOUSE GAS

Since biochar promotes soil fertility while storing carbon in the soil for hundreds or even thousands of years, using it as a soil supplement reduces greenhouse gas emissions both directly and indirectly. Improved soil fertility encourages plant development, which increases CO_2 uptake. The requirement for fertilizer input is decreased by improved soil fertility as well, which lowers carbon emissions from fertilizer production, transportation, and application (Laird et al., 2009).

4.9.2 ENHANCE SOIL QUALITY

When biochar is applied as a soil fertilizer, an improvement in soil fertility, soil microbial activity, and nutrient retention can be attributed to an increase in agricultural output (Kookana et al., 2011). Biochar can provide nutrients to the soil directly because it contains nutrients like N, P, and K. In addition, biochar indirectly enhances soil fertility by improving the soil's capacity to hold onto nutrients. By limiting nutrient losses from the soil, biochar also lowers the likelihood of surface water bodies becoming eutrophication (Sohi et al., 2010). Crop nutrient uptake increased on soil treated with biochar in the Amazon region because of improved soil nitrogen retention (Steiner et al., 2008).

4.9.3 IMPACT OF NANO BIOCHAR SOIL MICROBES

Researchers from all around the world have presented and suggested using biochar in many areas (Feng et al., 2021). The soil treated with biochar is known to improve soil qualities, including enhanced fertility and other advantages (Silber et al., 2010). Several studies have shown that, after being released into the natural environment, bulk biochar can transform into nano biochar depending on the pyrolysis temperature used (Liu et al., 2020a). Biological soil parameters that are strongly controlled by a wide range of microbial populations may be significantly impacted by the interaction of nanobiochar with soil. The relationship may regulate various microbial activities, which would impact the way soil behaves. The mobilization of macronutrients like phosphorus is said to be increased by the interactions of micro biochar with soil minerals (Mahmoud et al., 2020). The population of soil microorganisms may fluctuate due to variations in soil nutrients

4.9.4 NANOBIOCHAR APPLICATION IN CATALYSIS, AND ENERGY

Nanobiochar has been utilized as a sustainable agent in a variety of disciplines, including energy, catalysis, biomedicines, and fuel, in addition to applications in agriculture and the environment. Nanobiochar enhancers are preferred nanometal additions when combined with gasoline; they significantly reduce harmful emissions from automobiles. A study showed that using sugarcane nanobiochar, derived from bagasse, as an alternative to metal-based fuel additives yielded positive outcomes (Safieddin Ardebili et al., 2020). The qualities of biochar-based nanocomposites frequently improved significantly, including pore characteristics, active sites, functional groups, and catalytic destruction capacity; separation also became straightforward and quick (Ramadan et al., 2020). Moreover, nanobiochar and biochar nanocomposites have been thoroughly investigated for their efficacy as acid catalysts and enzyme support. By increasing surface area and loading capacity while lowering mass transfer resistance, nanobiochar-based biocatalytic devices increased the factors determining efficiency (Jenie et al., 2020). Nanobiochar has been utilized as a sensing material in a variety of electrochemical sensors (Yao et al., 2021). The benefits and drawbacks of biochar and nanobiochar are mentioned in Table 4.2.

TABLE 4.2 Benefits and Drawbacks of Various Biochar and Nanobiochar

Applications	Advantages	Disadvantages	References
Soil amendment	Retain water and nutrient, sustainable resources, reduce greenhouse gases, fertilizer consumption and low-cost.	PAHs contaminants and possibly heavy metals	Lehmann et al. (2011)
Storage material	High recyclability, sustainable resources, and low cost	Surface treatment require	Hao et al. (2013)
Fuel cell	Renewable fuel compared with coal	Low voltage, low power output and High ash content	Ahn et al. (2013)
Catalyst	Co-catalyst, low-cost	Low abrasive and relative low efficiency	Zhang et al. (2013)
Sorbent of contaminant	Low-cost, abundant, and sustainable resources,	Effectiveness of organic/inorganic	Houben et al. (2013)

4.9.5 AGRICULTURAL APPLICATION

Nanobiochar has many applications in the field of agriculture, according to several studies. Many agricultural uses of nano biochar include chemical remediation, microbial growth support, and disease management. The soil benefits from microorganisms in numerous ways, mostly due to the presence of various nutrients and ash. The health of the soil, cycling of nutrients, disease prevention, fertility, and crop production are all maintained by microbial communities, which are crucial to the functioning of the plant-soil ecosystem (Thies, 2015).

The use of nano biochar considerably improved the biomass and microbiological diversity of actinobacteria and bacteroidetes in soil contaminated with cadmium while significantly reducing the variety of proteobacteria, which predominated in polluted areas (Liu et al., 2020a). As an alternative, biochar has also been shown to be extremely effective at sorbing and sequestering pesticides due to its higher porosity, which created more adsorption sites than unamended soil. It may also reduce the desorption, leaching, mobility, and bioavailability of pesticides in soil layers, leading to the remediation of pesticide-contaminated soils (Safaei Khorram et al., 2016). Due to their increased affinity for pesticides, these carbon nanomaterials may be a viable option for pesticide restoration and a long-term solution to pesticide-caused soil pollution (Cernansky, 2015).

4.9.6 NANOBIOCHAR FOR REMOVAL OF CONTAMINANTS

The adsorption approach is frequently used to remove and reduce pollutants from water bodies due to the accessibility of a variety of plant-based sorbents such as biochar, clay minerals, resin, and activated carbon, simplicity, and effectiveness (Kumar et al., 2017). Nanobiochar made from rice husk biochar was able to remove 90% of fluoride from an aqueous solution (1 g L^{-1}) in about 60 minutes, which was one of the inorganic contaminants (Goswami & Kumar, 2018). The immediate sorption at several places with substantial nano biochar, as opposed to biochar, could be the cause of the higher removal of fluoride by nano biochar (Goswami et al., 2011). Tetracycline and Hg^{2+} were extracted from an aqueous solution using magnetic nano biochar. The interaction of hydrogen bonds, electrostatic interactions, and C-C interactions mostly controlled the elimination of tetracycline, whereas Hg^{2+} adsorption could be caused by the formation of Hg-C bonds. Electrostatic attraction pollutants have a variety of binding sites in nano biochar (Li et al., 2020a). The removal of heavy metal pollutants including arsenic, lead, copper, and chromium from biochar nanocomposites has now been demonstrated by multiple additional researchers, as indicated in Table 4.3.

TABLE 4.3 Removal of Pollutants by Adsorption with Biochar and Nanobiochar

Biochar material	Pyrolysis temperature oC	Composite type	Pollutant removed	Adsorption efficiency	References
Pinewood	600	(Magnetic biochar, hematite-modified biochar)	Arsenic (V)	265 mg/kg	Zhang et al. (2013)
Pinewood	600	Pinewood derived nanobiochar	Carbamazepine	95%	Naghdi et al. (2019b)
Sugarcane bagasse	550	MgO modified nanobiochar	Nitrate	95 mg/g	Zhang et al. (2012b)
Corn straw	600	MgO modified nanobiochar	Arsenic	14.36 mg/g	Yu et al. (2015)
Rice straw	550	Hydroxyapatite modified nanobiochar	Pb (II), Zn (II), Cu (II)	561.2 mg/g 925.9 mg/g 826.3 mg/g	Wang et al. (2018)
Corn Stover	600	ZnO/ ZnS modified nanobiochar	Cr (VI), Cu (II), Pb (II)	91.2 mg/g 24.5 mg/g 135 mg/g	Li et al. (2018)
Wheat Straw	600	Magnetic modified nanobiochar	Tetracycline and Hg (II)	268 mg/g, 127.4 mg/g	Li et al. (2020b)
Wheat Straw	600	Wheat straw-modified nanobiochar	Cadmium	P<0.05	Liu et al. (2020b)
Pinewood	600	Oxygen functionalized nanobiochar	CBZ	83%	Naghdi et al. (2017)
Corn Straw	600	Iron impregnated biochar	As (V)	BC- 0.017 mg/g FBC – 6.8 mg/g	Zhang et al. (2012a)

4.10 CONCLUSION AND FUTURE ASPECTS

Most of the studies have shown the importance of nano biochar or biochar nanocomposite used in agricultural and environmental problems. Heavy metals, dyes, organic and inorganic contaminants can all be effectively absorbed by nano biochar. Additionally, nano biochar has numerous uses, including significantly increasing plant growth and yields, nutrient recycling, fertilizer characteristics, soil supplements, and removing pathogenic effects, among others. This phenomenon is referred to as "the biochar effect" (Elad, 2012). Current innovations in biochar production, properties, application, and analysis methods are addressed. In general, the usage of char as high-value sustainable goods appears to have great potential. The qualities of biochar must be further enhanced and tailored for the right uses, though, to be successfully used for a variety of purposes. To achieve biochar potential as a soil amendment, sorbent, catalyst, standards, and methodologies used to identify its structure and qualities need to be improved. While biochar can act as a nanocarrier for the distribution of fertilizer and pesticides to plants and crops, the synthesis of application-based nano biochar involved the utilization of ball milling, sonication, carbonization, and grinding techniques. Each technique has specific benefits and drawbacks for producing nano biochar. The application of nano biochar showed significant environmental advantages, such as reducing organic contaminants in water bodies, playing a major role in waste management, reducing soil erosion, and assisting in the removal of heavy metals from water bodies. Hence, research will concentrate on the optimization of the production methods for increasing the yield of nano biochar, producing electrode supercapacitors, minimizing agglomeration, and developing new properties and information for future applications.

KEYWORDS

- **adsorption**
- **biochar**
- **nanobiochar**
- **nanocomposite**
- **plant growth**

REFERENCES

Ahn, S. Y., Eom, S. Y., Rhie, Y. H., Sung, Y. M., Moon, C. E., Choi, G. M., & Kim, D. J., (2013). Utilization of wood biomass char in a direct carbon fuel cell (DCFC) system. *Applied Energy*, *105*, 207–216. https://doi.org/10.1016/j.apenergy.2013.01.023.

An, X., Wu, Z., Shi, W., Qi, H., Zhang, L., Xu, X., & Yu, B., (2021). Biochar for simultaneously enhancing the slow-release performance of fertilizers and minimizing the pollution of pesticides. *Journal of Hazardous Materials*, *407*, 124865. https://doi.org/10.1016/j.jhazmat.2020.124865.

Ashiq, A., Adassooriya, N. M., Sarkar, B., Rajapaksha, A. U., Ok, Y. S., & Vithanage, M., (2019). Municipal solid waste biochar-bentonite composite for the removal of antibiotic ciprofloxacin from aqueous media. *Journal of Environmental Management, 236*, 428–435. https://doi.org/10.1016/j.jenvman.2019.02.006.

Awad, Y. M., Lee, S. S., Kim, K. H., Ok, Y. S., & Kuzyakov, Y., (2018). Carbon and nitrogen mineralization and enzyme activities in soil aggregate-size classes: Effects of biochar, oyster shells, and polymers. *Chemosphere, 198*, 40–48. https://doi.org/10.1016/j.chemosphere.2018.01.034.

Bargmann, I., Rillig, M. C., Buss, W., Kruse, A., & Kuecke, M., (2013). Hydrochar and biochar effects on germination of spring barley. *Journal of Agronomy and Crop Science, 199*(5), 360–373. https://doi.org/10.1111/jac.12024.

Bi, S., Barinelli, V., & Sobkowicz, M. J., (2020). Degradable controlled release fertilizer composite prepared via extrusion: Fabrication, characterization, and release mechanisms. *Polymers, 12*(2), 301. https://doi.org/10.3390/polym12020301.

Bolan, N. S., Thangarajan, R., Seshadri, B., Jena, U., Das, K. C., Wang, H., & Naidu, R., (2013). Landfills as a biorefinery to produce biomass and capture biogas. *Bioresource Technology, 135*, 578–587. https://doi.org/10.1016/j.biortech.2012.08.135.

Brewer, C. E., Chuang, V. J., Masiello, C. A., Gonnermann, H., Gao, X., Dugan, B., Driver, L. E., et al., (2014). New approaches to measuring biochar density and porosity. *Biomass and Bioenergy, 66*, 176–185. https://doi.org/10.1016/j.biombioe.2014.03.059.

Cernansky, R., (2015). Agriculture: State-of-the-art soil. *Nature, 517*(7534), 258–260. https://doi.org/10.1038/517258a.

Chausali, N., Saxena, J., & Prasad, R., (2021). Nanobiochar and biochar based nanocomposites: Advances and applications. *Journal of Agriculture and Food Research, 5*, 100191. https://doi.org/10.1016/j.jafr.2021.100191.

Chen, B., Chen, Z., & Lv, S., (2011). A novel magnetic biochar efficiently sorbs organic pollutants and phosphate. *Bioresource Technology, 102*(2), 716–723. https://doi.org/10.1016/j.biortech.2010.08.067.

Chen, L., Chen, X. L., Zhou, C. H., Yang, H. M., Ji, S. F., Tong, D. S., Zhong, Z. K., et al., (2017). Environmental-friendly montmorillonite-biochar composites: Facile production and tunable adsorption-release of ammonium and phosphate. *Journal of Cleaner Production, 156*, 648–659. https://doi.org/10.1016/j.jclepro.2017.04.050.

Da Costa, T. P., Westphalen, G., Nora, F. B. D., De Zorzi, S. B., & Rosa, G. S. Da., (2019). Technical and environmental assessment of coated urea production with a natural polymeric suspension in spouted bed to reduce nitrogen losses. *Journal of Cleaner Production, 222*, 324–334. https://doi.org/10.1016/j.jclepro.2019.03.069.

Elad, Y., Cytryn, E., Harel, Y. M., Lew, B., & Graber, E. R., (2011). The biochar effect: Plant resistance to biotic stresses. *Phytopathologia Mediterranea, 50*(3), 335–349.

El-Naggar, A., Lee, S. S., Rinklebe, J., Farooq, M., Song, H., Sarmah, A. K., Zimmerman, A. R., et al., (2019). Biochar application to low fertility soils: A review of current status, and future prospects. *Geoderma, 337*, 536–554. https://doi.org/10.1016/j.geoderma.2018.09.034.

Feng, W., Yang, F., Cen, R., Liu, J., Qu, Z., Miao, Q., & Chen, H., (2021). Effects of straw biochar application on soil temperature, available nitrogen and growth of corn. *Journal of Environmental Management, 277*, 111331. https://doi.org/10.1016/j.jenvman.2020.111331.

Föhr, J., Ranta, T., Suikki, J., & Soininen, H., (2017). Manufacturing of torrefied pellets without a binder from different raw wood materials in the pilot plant. *Wood Res., 62*(3), 481–494.

Gai, X., Wang, H., Liu, J., Zhai, L., Liu, S., Ren, T., & Liu, H., (2014). Effects of feedstock and pyrolysis temperature on biochar adsorption of ammonium and nitrate. *PLoS One*, *9*(12), e113888. https://doi.org/10.1371/journal.pone.0113888.

García-Sánchez, F., Simón-Grao, S., Martínez-Nicolás, J. J., Alfosea-Simón, M., Liu, C., Chatzissavvidis, C., Pérez-Pérez, J. G., & Cámara-Zapata, J. M., (2020). Multiple stresses occurring with boron toxicity and deficiency in plants. *Journal of Hazardous Materials*, *397*, 122713. https://doi.org/10.1016/j.jhazmat.2020.122713.

Goswami, R., & Kumar, M., (2018). Removal of fluoride from aqueous solution using nanoscale rice husk biochar. *Groundwater for Sustainable Development*, *7*, 446–451. https://doi.org/10.1016/j.gsd.2017.12.010.

Goswami, R., Deb, P., Thakur, R., Sarma, K. P., & Basumallick, A., (2011). Removal of As(III) from aqueous solution using functionalized ultrafine iron oxide nanoparticles. *Separation Science and Technology*, *46*(6), 1017–1022. https://doi.org/10.1080/01496395.2010.537728.

Guo, F., Bao, L., Wang, H., Larson, S. L., Ballard, J. H., Knotek-Smith, H. M., Zhang, Q., et al., (2020). A simple method for the synthesis of biochar nanodots using hydrothermal reactor. *MethodsX*, *7*, 101022. https://doi.org/10.1016/j.mex.2020.101022.

Gupta, B., Mishra, A., Singh, R., & Thakur, I. S., (2021). Fabrication of calcite based biocomposites for catalytic removal of heavy metals from electroplating industrial effluent. *Environmental Technology & Innovation*, *21*, 101278. https://doi.org/10.1016/j.eti.2020.101278.

Gurwick, N. P., Moore, L. A., Kelly, C., & Elias, P., (2013). A systematic review of biochar research, with a focus on its stability in situ and its promise as a climate mitigation strategy. *PLoS One*, *8*(9), e75932. https://doi.org/10.1371/journal.pone.0075932.

Hao, W., Björkman, E., Lilliestråle, M., & Hedin, N., (2013). Activated carbons prepared from hydrothermally carbonized waste biomass used as adsorbents for CO_2. *Applied Energy*, *112*, 526–532. https://doi.org/10.1016/j.apenergy.2013.02.028.

Houben, D., Evrard, L., & Sonnet, P., (2013). Mobility, bioavailability and pH-dependent leaching of cadmium, zinc and lead in a contaminated soil amended with biochar. *Chemosphere*, *92*(11), 1450–1457. https://doi.org/10.1016/j.chemosphere.2013.03.055.

Huang, X., Zhu, S., Zhang, H., Huang, Y., Wang, X., Wang, Y., & Chen, D., (2021). Biochar nanoparticles induced distinct biological effects on freshwater algae via oxidative stress, membrane damage, and nutrient depletion. *ACS Sustainable Chemistry & Engineering*, *9*(32), 10761–10770. https://doi.org/10.1021/acssuschemeng.1c01814.

Igalavithana, A. D., Ok, Y. S., Usman, A. R. A., Al-Wabel, M. I., Oleszczuk, P., & Lee, S. S., (2015). *The Effects of Biochar Amendment on Soil Fertility* (pp. 123–144). https://doi.org/10.2136/sssaspecpub63.2014.0040.

Jenie, S. N. A., Kristiani, A., Sudiyarmanto, Khaerudini, D. S., & Takeishi, K., (2020). Sulfonated magnetic nanobiochar as heterogeneous acid catalyst for esterification reaction. *Journal of Environmental Chemical Engineering*, *8*(4), 103912. https://doi.org/10.1016/j.jece.2020.103912.

Jiang, C., Bo, J., Xiao, X., Zhang, S., Wang, Z., Yan, G., Wu, Y., Wong, C., & He, H., (2020). Converting waste lignin into nano-biochar as a renewable substitute of carbon black for reinforcing styrene-butadiene rubber. *Waste Management*, *102*, 732–742. https://doi.org/10.1016/j.wasman.2019.11.019.

Kambo, H. S., & Dutta, A., (2015). A comparative review of biochar and hydrochar in terms of production, physico-chemical properties and applications. *Renewable and Sustainable Energy Reviews*, *45*, 359–378. https://doi.org/10.1016/j.rser.2015.01.050.

Kookana, R. S., Sarmah, A. K., Van, Z. L., Krull, E., & Singh, B., (2011). *Biochar Application to Soil* (pp. 103–143). https://doi.org/10.1016/B978-0-12-385538-1.00003-2.

Kumar, M., Xiong, X., Wan, Z., Sun, Y., Tsang, D. C. W., Gupta, J., Gao, B., et al., (2020). Ball milling as a mechanochemical technology for fabrication of novel biochar nanomaterials. *Bioresource Technology*, *312*, 123613. https://doi.org/10.1016/j.biortech.2020.123613.

Kumar, P., Saraswat, C., Mishra, B. K., Avtar, R., Patel, H., Patel, A., Sharma, T., & Patel, R., (2017). Batch technique to evaluate the efficiency of different natural adsorbents for defluoridation from groundwater. *Applied Water Science*, *7*(5), 2597–2606. https://doi.org/10.1007/s13201-016-0473-5.

Laird, D. A., Brown, R. C., Amonette, J. E., & Lehmann, J., (2009). Review of the pyrolysis platform for coproducing bio-oil and biochar. *Biofuels, Bioproducts and Biorefining*, *3*(5), 547–562. https://doi.org/10.1002/bbb.169.

Lehmann, J. J. S., (2012). In: Lehmann, J., & Joseph, S., (eds.), *Biochar for Environmental Management*. Routledge. https://doi.org/10.4324/9781849770552.

Lehmann, J., & Joseph, S., (2015). *Biochar for Environmental Management: Science, Technology, and Implementation*. Routledge.

Lehmann, J., (2007). Bioenergy in the black. *Front. Ecol. Environ.*

Lehmann, J., Rillig, M. C., Thies, J., Masiello, C. A., Hockaday, W. C., & Crowley, D., (2011). Biochar effects on soil biota – A review. *Soil Biology and Biochemistry*, *43*(9), 1812–1836. https://doi.org/10.1016/j.soilbio.2011.04.022.

Leng, L., Yuan, X., Zeng, G., Shao, J., Chen, X., Wu, Z., Wang, H., & Peng, X., (2015). Surface characterization of rice husk bio-char produced by liquefaction and application for cationic dye (malachite green) adsorption. *Fuel*, *155*, 77–85. https://doi.org/10.1016/j.fuel.2015.04.019.

Li, C., Zhang, L., Gao, Y., & Li, A., (2018). Facile synthesis of nano ZnO/ZnS modified biochar by directly pyrolyzing of zinc contaminated corn stover for Pb(II), Cu(II) and Cr(VI) removals. *Waste Management*, *79*, 625–637. https://doi.org/10.1016/j.wasman.2018.08.035.

Li, R., Zhang, Y., Deng, H., Zhang, Z., Wang, J. J., Shaheen, S. M., Xiao, R., et al., (2020a). Removing tetracycline and Hg(II) with ball-milled magnetic nanobiochar and its potential on polluted irrigation water reclamation. *Journal of Hazardous Materials*, *384*, 121095. https://doi.org/10.1016/j.jhazmat.2019.121095.

Li, R., Zhang, Y., Deng, H., Zhang, Z., Wang, J. J., Shaheen, S. M., Xiao, R., et al., (2020b). Removing tetracycline and Hg(II) with ball-milled magnetic nanobiochar and its potential on polluted irrigation water reclamation. *Journal of Hazardous Materials*, *384*, 121095. https://doi.org/10.1016/j.jhazmat.2019.121095.

Li, W., Zhu, X., He, Y., Xing, B., Xu, J., & Brookes, P. C., (2013). Enhancement of water solubility and mobility of phenanthrene by natural soil nanoparticles. *Environmental Pollution*, *176*, 228–233. https://doi.org/10.1016/j.envpol.2013.01.011.

Libra, J. A., Ro, K. S., Kammann, C., Funke, A., Berge, N. D., Neubauer, Y., Titirici, M. M., et al., (2011). Hydrothermal carbonization of biomass residuals: A comparative review of the chemistry, processes and applications of wet and dry pyrolysis. *Biofuels*, *2*(1), 71–106. https://doi.org/10.4155/bfs.10.81.

Liu, G., Zheng, H., Jiang, Z., Zhao, J., Wang, Z., Pan, B., & Xing, B., (2018). Formation and physicochemical characteristics of nano biochar: Insight into chemical and colloidal stability. *Environmental Science & Technology*, *52*(18), 10369–10379. https://doi.org/10.1021/acs.est.8b01481.

Liu, W., Li, Y., Feng, Y., Qiao, J., Zhao, H., Xie, J., Fang, Y., et al., (2020a). The effectiveness of nanobiochar for reducing phytotoxicity and improving soil remediation in cadmium-contaminated soil. *Scientific Reports*, *10*(1), 858. https://doi.org/10.1038/s41598-020-57954-3.

Liu, W., Li, Y., Feng, Y., Qiao, J., Zhao, H., Xie, J., Fang, Y., et al., (2020b). The effectiveness of nanobiochar for reducing phytotoxicity and improving soil remediation in cadmium-contaminated soil. *Scientific Reports*, *10*(1), 858. https://doi.org/10.1038/s41598-020-57954-3.

Lyu, H., Gao, B., He, F., Zimmerman, A. R., Ding, C., Huang, H., & Tang, J., (2018). Effects of ball milling on the physicochemical and sorptive properties of biochar: Experimental observations and governing mechanisms. *Environmental Pollution*, *233*, 54–63. https://doi.org/10.1016/j.envpol.2017.10.037.

Mahmoud, E., El Baroudy, A., Ali, N., & Sleem, M., (2020). Spectroscopic studies on the phosphorus adsorption in salt-affected soils with or without nano-biochar additions. *Environmental Research*, *184*, 109277. https://doi.org/10.1016/j.envres.2020.109277.

Makshut, N. A., Ngaini, Z., Wahi, R., Hussain, H., Mahmut, N. I., & Bahrin, N. Q., (2020). Nano-sized adsorbent from pyrolyzed sago activated sludge for removal of Pb(II) from aqueous solution. *Pertanika Journal of Science and Technology*, *28*(3).

Mishra, A., Gupta, B., Kumar, N., Singh, R., Varma, A., & Thakur, I. S., (2020). Synthesis of calcite-based bio-composite biochar for enhanced biosorption and detoxification of chromium Cr (VI) by *Zhihengliuella* sp. ISTPL4. *Bioresource Technology*, *307*, 123262. https://doi.org/10.1016/j.biortech.2020.123262.

Mohan, D., Pittman, C. U., & Steele, P. H., (2006). Pyrolysis of wood/biomass for bio-oil: A critical review. *Energy & Fuels*, *20*(3), 848–889. https://doi.org/10.1021/ef0502397.

Naghdi, M., Taheran, M., Brar, S. K., Kermanshahi-pour, A., Verma, M., & Surampalli, R. Y., (2017). Immobilized laccase on oxygen functionalized nanobiochars through mineral acids treatment for removal of carbamazepine. *Science of The Total Environment*, *584, 585*, 393–401. https://doi.org/10.1016/j.scitotenv.2017.01.021.

Naghdi, M., Taheran, M., Brar, S. K., Kermanshahi-pour, A., Verma, M., & Surampalli, R. Y., (2018). Pinewood nanobiochar: A unique carrier for the immobilization of crude laccase by covalent bonding. *International Journal of Biological Macromolecules*, *115*, 563–571. https://doi.org/10.1016/j.ijbiomac.2018.04.105.

Naghdi, M., Taheran, M., Pulicharla, R., Rouissi, T., Brar, S. K., Verma, M., & Surampalli, R. Y., (2019a). Pine-wood derived nanobiochar for removal of carbamazepine from aqueous media: Adsorption behavior and influential parameters. *Arabian Journal of Chemistry*, *12*(8), 5292–5301. https://doi.org/10.1016/j.arabjc.2016.12.025.

Naghdi, M., Taheran, M., Pulicharla, R., Rouissi, T., Brar, S. K., Verma, M., & Surampalli, R. Y., (2019b). Pine-wood derived nanobiochar for removal of carbamazepine from aqueous media: Adsorption behavior and influential parameters. *Arabian Journal of Chemistry*, *12*(8), 5292–5301. https://doi.org/10.1016/j.arabjc.2016.12.025.

Noreen, S., & Abd-Elsalam, K. A., (2021). Biochar-based nanocomposites: A sustainable tool in wastewater bioremediation. In *Aquananotechnology* (pp. 185–200). Elsevier. https://doi.org/10.1016/B978-0-12-821141-0.00023-9.

Oleszczuk, P., Ćwikła-Bundyra, W., Bogusz, A., Skwarek, E., & Ok, Y. S., (2016). Characterization of nanoparticles of biochars from different biomass. *Journal of Analytical and Applied Pyrolysis*, *121*, 165–172. https://doi.org/10.1016/j.jaap.2016.07.017.

Pang, W., Hou, D., Wang, H., Sai, S., Wang, B., Ke, J., Wu, G., et al., (2018). Preparation of microcapsules of slow-release NPK compound fertilizer and the release characteristics. *Journal of the Brazilian Chemical Society*. https://doi.org/10.21577/0103-5053.20180117.

Ramadan, M. M., Asran-Amal, & Abd-Elsalam, K. A., (2020). Micro/nano biochar for sustainable plant health: Present status and future prospects. In: *Carbon Nanomaterials for Agri-Food and Environmental Applications* (pp. 323–357). Elsevier. https://doi.org/10.1016/B978-0-12-819786-8.00016-5.

Ramanayaka, S., Tsang, D. C. W., Hou, D., Ok, Y. S., & Vithanage, M., (2020). Green synthesis of graphitic nanobiochar for the removal of emerging contaminants in aqueous media. *Science of The Total Environment*, *706*, 135725. https://doi.org/10.1016/j.scitotenv.2019.135725.

Ramanayaka, S., Vithanage, M., Alessi, D. S., Liu, W. J., Jayasundera, A. C. A., & Ok, Y. S., (2020). Nanobiochar: Production, properties, and multifunctional applications. *Environmental Science: Nano*, *7*(11), 3279–3302. https://doi.org/10.1039/D0EN00486C.

Read, D. J., Leake, J. R., & Perez-Moreno, J., (2004). Mycorrhizal fungi as drivers of ecosystem processes in heathland and boreal forest biomes. *Canadian Journal of Botany*, *82*(8), 1243–1263. https://doi.org/10.1139/b04-123.

Safaei, K. M., Zhang, Q., Lin, D., Zheng, Y., Fang, H., & Yu, Y., (2016). Biochar: A review of its impact on pesticide behavior in soil environments and its potential applications. *Journal of Environmental Sciences*, *44*, 269–279. https://doi.org/10.1016/j.jes.2015.12.027.

Safieddin, A. S. M., Taghipoor, A., Solmaz, H., & Mostafaei, M., (2020). The effect of nano-biochar on the performance and emissions of a diesel engine fueled with fusel oil-diesel fuel. *Fuel*, *268*, 117356. https://doi.org/10.1016/j.fuel.2020.117356.

Sajjadi, B., Chen, W. Y., Mattern, D. L., Hammer, N., & Dorris, A., (2020). Low-temperature acoustic-based activation of biochar for enhanced removal of heavy metals. *Journal of Water Process Engineering*, *34*, 101166. https://doi.org/10.1016/j.jwpe.2020.101166.

Serrà, A., Zhang, Y., Sepúlveda, B., Gómez, E., Nogués, J., Michler, J., & Philippe, L., (2020). Highly reduced ecotoxicity of ZnO-based micro/nanostructures on aquatic biota: Influence of architecture, chemical composition, fixation, and photocatalytic efficiency. *Water Research*, *169*, 115210. https://doi.org/10.1016/j.watres.2019.115210.

Silber, A., Levkovitch, I., & Graber, E. R., (2010). pH-dependent mineral release and surface properties of cornstraw biochar: Agronomic implications. *Environmental Science & Technology*, *44*(24), 9318–9323. https://doi.org/10.1021/es101283d.

Sohi, S. P., Krull, E., Lopez-Capel, E., & Bol, R., (2010). *A Review of Biochar and Its Use and Function in Soil* (pp. 47–82). https://doi.org/10.1016/S0065-2113(10)05002-9.

Steiner, C., Glaser, B., Geraldes, T. W., Lehmann, J., Blum, W. E. H., & Zech, W., (2008). Nitrogen retention and plant uptake on a highly weathered central Amazonian Ferralsol amended with compost and charcoal. *Journal of Plant Nutrition and Soil Science*, *171*(6), 893–899. https://doi.org/10.1002/jpln.200625199.

Sun, Y., Lyu, H., Cheng, Z., Wang, Y., & Tang, J., (2022). Insight into the mechanisms of ball-milled biochar addition on soil tetracycline degradation enhancement: Physicochemical properties and microbial community structure. *Chemosphere*, *291*, 132691. https://doi.org/10.1016/j.chemosphere.2021.132691.

Taheran, M., Naghdi, M., Brar, S. K., Knystautas, E. J., Verma, M., Ramirez, A. A., Surampalli, R. Y., & Valero, J. R., (2016). Adsorption study of environmentally relevant concentrations of chlortetracycline on pinewood biochar. *Science of The Total Environment, 571*, 772–777. https://doi.org/10.1016/j.scitotenv.2016.07.050.

Thies, J. E., Rillig, M. C., & Graber, E. R., (2015). Biochar effects on the abundance, activity, and diversity of the soil biota. In: *Biochar for Environmental Management* (pp. 327–389). Routledge.

Torney, F., Trewyn, B. G., Lin, V. S. Y., & Wang, K., (2007). Mesoporous silica nanoparticles deliver DNA and chemicals into plants. *Nature Nanotechnology*, *2*(5), 295–300. https://doi.org/10.1038/nnano.2007.108.

Uchimiya, M., Chang, S., & Klasson, K. T., (2011). Screening biochars for heavy metal retention in soil: Role of oxygen functional groups. *Journal of Hazardous Materials*, *190*(1–3), 432–441. https://doi.org/10.1016/j.jhazmat.2011.03.063.

Varma, A. K., Shankar, R., & Mondal, P., (2018). A review on pyrolysis of biomass and the impacts of operating conditions on product yield, quality, and upgradation. In: *Recent Advancements in Biofuels and Bioenergy Utilization* (pp. 227–259). Springer Singapore. https://doi.org/10.1007/978-981-13-1307-3_10.

Wang, Y. Y., Liu, Y. X., Lu, H. H., Yang, R. Q., & Yang, S. M., (2018). Competitive adsorption of Pb(II), Cu(II), and Zn(II) ions onto hydroxyapatite-biochar nanocomposite in aqueous solutions. *Journal of Solid-State Chemistry*, *261*, 53–61. https://doi.org/10.1016/j.jssc.2018.02.010.

Wang, Y., Wang, L., Li, Z., Yang, D., Xu, J., & Liu, X., (2021). MgO-laden biochar enhances the immobilization of Cd/Pb in aqueous solution and contaminated soil. *Biochar*, *3*(2), 175–188. https://doi.org/10.1007/s42773-020-00080-0.

Weber, K., & Quicker, P., (2018). Properties of biochar. *Fuel*, *217*, 240–261. https://doi.org/10.1016/j.fuel.2017.12.054.

Xu, X., & Cölfen, H., (2021). Ultracentrifugation techniques for the ordering of nanoparticles. *Nanomaterials*, *11*(2), 333. https://doi.org/10.3390/nano11020333.

Yaashikaa, P. R., Kumar, P. S., Varjani, S., & Saravanan, A., (2020). A critical review on the biochar production techniques, characterization, stability and applications for circular bioeconomy. *Biotechnology Reports*, *28*, e00570. https://doi.org/10.1016/j.btre.2020.e00570.

Yao, L., He, L., Yang, Y., Zhang, Y., Liu, Z., Liang, L., & Piao, Y., (2021). Nanobiochar paper based electrochemical immunosensor for fast and ultrasensitive detection of microcystin-LR. *Science of The Total Environment*, *750*, 141692. https://doi.org/10.1016/j.scitotenv.2020.141692.

Yu, Z., Zhou, L., Huang, Y., Song, Z., & Qiu, W., (2015). Effects of a manganese oxide-modified biochar composite on adsorption of arsenic in red soil. *Journal of Environmental Management*, *163*, 155–162. https://doi.org/10.1016/j.jenvman.2015.08.020.

Yuan, P., Shen, B., Duan, D., Adwek, G., Mei, X., & Lu, F., (2017). Study on the formation of direct reduced iron by using biomass as reductants of carbon containing pellets in RHF process. *Energy*, *141*, 472–482. https://doi.org/10.1016/j.energy.2017.09.058.

Yuan, Y., Zhang, N., & Hu, X., (2020). Effects of wet and dry ball milling on the physicochemical properties of sawdust derived-biochar. *Instrumentation Science & Technology*, *48*(3), 287–300. https://doi.org/10.1080/10739149.2019.1708751.

Zhang, M., Gao, B., Varnoosfaderani, S., Hebard, A., Yao, Y., & Inyang, M., (2013). Preparation and characterization of a novel magnetic biochar for arsenic removal. *Bioresource Technology*, *130*, 457–462. https://doi.org/10.1016/j.biortech.2012.11.132.

Zhang, M., Gao, B., Yao, Y., Xue, Y., & Inyang, M., (2012a). Synthesis, characterization, and environmental implications of graphene-coated biochar. *Science of The Total Environment*, *435–436*, 567–572. https://doi.org/10.1016/j.scitotenv.2012.07.038.

Zhang, M., Gao, B., Yao, Y., Xue, Y., & Inyang, M., (2012b). Synthesis of porous MgO-biochar nanocomposites for removal of phosphate and nitrate from aqueous solutions. *Chemical Engineering Journal*, *210*, 26–32. https://doi.org/10.1016/j.cej.2012.08.052.

Zhang, S., Asadullah, M., Dong, L., Tay, H. L., & Li, C. Z., (2013). An advanced biomass gasification technology with integrated catalytic hot gas cleaning. Part II: Tar reforming using char as a catalyst or as a catalyst support. *Fuel*, *112*, 646–653. https://doi.org/10.1016/j.fuel.2013.03.015.

Zhang, X., Wells, M., Niazi, N. K., Bolan, N., Shaheen, S., Hou, D., Gao, B., et al., (2022). Nanobiochar-rhizosphere interactions: Implications for the remediation of heavy-metal contaminated soils. *Environmental Pollution*, *299*, 118810. https://doi.org/10.1016/j.envpol.2022.118810.

CHAPTER 5

Nano- and Nano-Biochar in Reducing Soil Stress: An Integrated Approach for Sustainable Agriculture

SEYEDE ROGHIE GHADIRNEZHAD SHIADE,[1] AMIN FATHI,[2] and REZA RAHIMI[3]

[1]*Department of Agronomy, Sari Agricultural Science and Natural Resources University, Sari, Iran*

[2]*Department of Agronomy, Ayatollah Amoli Branch, Islamic Azad University, Amol, Iran*

[3]*Department of Agronomy and Plant Breeding, Karaj Branch, Islamic Azad University, Karaj, Iran*

ABSTRACT

Due to the impairment of physiological, morphological, and biochemical functions in plants, salinity is a critical abiotic stress that diminishes plant yield and productivity. Furthermore, this abiotic stress can negatively affect various aspects of the environment. Recently, nanotechnology is increasingly being used in various fields, especially in agriculture, as it can provide quick and effective solutions that are indispensable to achieving sustainability. Accordingly, the advantages of biochar are paired with nanoparticles that exhibit enhanced physicochemical properties in new classes of biochar materials, such as nanobiochar. Because of the large specific surface area and strong cation exchange capacity of these compounds, especially when exposed to salinity, these materials have pronounced fertilization potential in agriculture and supply sufficient nutrients in soil. In this chapter, the causes, mechanisms, and various morphological, physiological, and biochemical

Nanomaterials and Nano-Biochar in Reducing Soil Stress: An Integrated Approach to Sustainable Agriculture. Vishnu D. Rajput, & Tatiana Minkina (Eds.)

impacts of salinity on plants were first delineated. In addition, the effects on the environment were mentioned. Furthermore, various solutions based on nanotechnology were presented and the types of nanoparticles (NPs), their applications, and their absorption by plants were explained. Moreover, nanobiochar and its use to mitigate the effects of salinity were discussed. Overall, it can be concluded that nanotechnology and nanobiochar could be introduced as emerging effective methods to combat the effects of salt stress on both plants and the environment.

5.1 INTRODUCTION

By 2050, nearly 10 billion people will live in the world, representing a 50 percent increase in food demand (Zulfiqar & Ashraf, 2021). In the face of climate change and an ever-growing population, current agricultural practices appear to be unsustainable (Rajput et al., 2022). Globally, one-third of arable land suffers from salinization as a result of poor irrigation practices, improper use of agrochemicals, and industrial pollution (Zulfiqar & Ashraf, 2021). In other words, plant production is threatened by salinity, a major abiotic stress that poses an important challenge to food and feed security. Plant growth can be repressed by salts in soil for two reasons. As a first consequence, salt elements in the soil solution shrink the plant's capacity to absorb water, resulting in a reduction in growth rate. The effect described above is called osmotic pressure or water deficit in salinity. Salt can also impair cells in transpiring leaves when additional amounts of saline elements enter the plant through the transpiration flow, which is called the salt-specific effect or ion excess of salinity (Parihar et al., 2015). The effect of salinity is threefold: it diminishes water potential, leads to ion imbalance, and disrupts ion homeostasis (Parihar et al., 2015). Consequently, photosynthetic activity, stomatal conductivity, gas exchange capacity, and electron transport in cellular organelles can be impacted, which in turn leads to a great amount of reactive oxygen species (ROS) generation such as superoxide radicals, singlet oxygen, hydroxyl radicals, and hydrogen peroxides (Liang et al., 2018). Plant cells' lipid membranes, proteins, nucleic acids, and other macromolecules are severely damaged by ROS, leading to oxidative stress and cell death (Farhangi-Abriz & Ghassemi-Golezani, 2021).

The use of solutions based on nanotechnology is rapidly increasing in various fields. Agricultural nanotechnology has gained considerable ground in recent years as a rapidly developing area (Zulfiqar & Ashraf, 2021) because it provides potential and rapid solutions crucial to sustainability (Hofmann et al., 2020). For example, applications of nanoparticles (nano-fertilizers, nano-pesticides), certain nanomaterials (e.g., carbon-based single- and multi-walled nanotubes, graphene, fullerene, fullerol, and polymeric chitosan), and nano-based methods (nanofiltration of brackish and/or gray water resources for irrigated cultivation, transport of trace elements and their deposition in plant tissues) have emerged as promising alternatives to mitigate nutrient perturbations and improve plant food production under various abiotic conditions, such as severe salinity and induced salt perturbations (Ondrasek et al., 2022). According to Pandey et al. (2018), NaCl-exposed (100 mM) growth medium treated with graphene or carbon-based nanotubes significantly reduced salt stress symptoms in treated plants. The results can be attributed to (a) the effect of carbon-based nanoparticles on plant transcriptome performance (such as increased expression of aquaporins) and (b) the physical connections between carbon-based nanoparticles and toxic ions (such as the removal of Na^+ from salt solutions). In another study, multi-walled carbon nanotubes were found to boost the salt tolerance of oilseed rape seedlings treated with NaCl stress (Zhao et al., 2019). Furthermore, phytonutrients can also be combined into naturally occurring zeolites, namely Si-Al minerals with a high active surface area and a 10-fold higher cation exchange capacity (CEC) (Ditta & Arshad, 2016). Additionally, Gohari et al. (2020) identified that titanium dioxide nanoparticles (TiO_2 NPs) improve growth and amend salinity stress influences on essential oil compounds and biochemical characteristics of *Dracocephalum moldavica*.

The term biochar refers to a carbonaceous material produced from bioenergy or agricultural feedstocks after pyrolysis (at low temperatures and under restricted oxygen circumstances) and is widely used to improve crop yields in bulk form (Ali et al., 2017; Ghadirnezhad Shiade et al., 2022a). Recently, however, the physical comminution of bulk biochar has enabled the production of nanoscale biochar with a size close to 100 nm (Chen et al., 2017), with the size varying contingent on the utilized feedstock and pyrolysis temperature (Wang et al., 2016). Over the past few years, much has been reported on the potential of nanobiochar to mediate salt tolerance in plants. Based on spectroscopic analysis, Mahmoud et al. (2020) noted that soil P adsorption is enhanced by nanobiochar under salt stress. Although several agronomic and non-agronomic approaches have been proposed to address the adverse effects of salinity, the use of nano- and nanobiochar are novel methods. Therefore, these approaches were presented in more detail in this chapter.

5.2 SALINITY IMPACTS ON PLANTS

5.2.1 WHAT IS SALINITY?

Salinity is defined as the condition in which soluble salts accumulate in a solution (water or soil) to such an extent that they impair the growth and development of plants (Shrivastava & Kumar, 2015; Negacz et al., 2022). Salt-affected soils are defined by changes occurring on their surface, such as the white salt crusts that appear on bare land and the salt stains that form in the soil; constant or persistent waterlogging after rainfall; deterioration of roads; surface and ground water deterioration, making it unsuitable for human and livestock use (Shrivastava & Kumar, 2015). The electrical conductivity of soil saturation extracts is widely used to measure soil salinity (ECe, in dS m^{-1}) (Negacz et al., 2022). Normal soils have an exchangeable sodium percentage (ESP) below 15, a pH range between 6.5 and 8.5, and a conductivity (EC) below 4.0 dS m^{-1}, as well as equivalent levels of anions and cations. Saline soils, on the other hand, have an ESP of about 15.0, sodium adsorption ratios (SARs) of not more than 13.0 (at 25°C), and ECs of >4.0 dS m^{-1} and pH < 8.5 (Mindari et al., 2018).

5.2.2 SALINITY CAUSES

Primary and secondary salinization are the two types of salinization. In the case of the former, several natural processes lead to the development of salinity, including weathering of parent material, deposition of sandstones, exposure of coastal areas and rivers to seawater, and evapotranspiration that leaves large amounts of salt in the soil. Precipitation causes ocean salts (e.g., sodium chloride) to accumulate in the soil as a result of cyclones (Chaitanya et al., 2014; Ruto et al., 2018). Furthermore, salinization could be caused by water seeping from seas into low-lying areas and arid zones with little water to leach out excess salts (Ruto et al., 2018). Notably, climate change is likely to exacerbate many of these processes. This environmental issue may occur in arid and semi-arid regions due to evaporation, where salts steadily accumulate. Additionally, saline soils can be blown inland by wind (Ruto et al., 2018). Secondary salinization occurs when soil in both upper and lower layers is moved by anthropogenic activities (Ruto et al., 2018). Saline fertilizers or irrigation without proper rainfall or drainage triggers soil and water salinization, making the soil unproductive due to salt accumulation in root zones after evaporation (Shiade & Boelt, 2020). Moreover, waterlogging and inadequate drainage also degrade its quality. As a result of poor irrigation practices, salinization continues to increase in new areas (Patel & Dave, 2011).

5.2.3 SALINITY MECHANISM

In soil, salinity affects physical properties primarily through a process known as flocculation, in which fine soil particles coalesce into aggregates. Although it promotes soil aeration and root growth, high salinity has a detrimental effect on plant growth and yield. Magnesium and calcium keep the soil flocculent, while sodium disperses it. The presence of salts causes the soil to become hard and cementitious, resulting in lower infiltration, lowered hydraulic conductivity (HC), and crusting on the surface (Srivastava et al., 2019).

Plants and microbes may be able to access water and mineral nutrients depending on the salt concentration in the soil solution. Consequently, soil organic matter, content, and stability are often affected by a high concentration of soluble salts (Mavi et al., 2012). Furthermore, increasing immobilization beyond the root reach could explain the low organic matter and mineral content of saline soils (Srivastava et al., 2019). Additionally, these soils contain increased levels of boron, aluminum, and carbonate elements affecting both community diversity and the number of microorganisms (Imadi et al., 2016). In other words, increased salinity negatively affects soil composition by increasing the metals, soluble chloride, sulfate, and bicarbonate ions concentration, whereas magnesium and calcium ions decrease (Abeer et al., 2014).

5.2.4 SALINITY CONSEQUENCE IN PLANTS

Salt stress affects crops differently. A combination of morphological, physiological, and biochemical processes leads to salinity impacts on seed germination, and water and nutrient absorption (Akbarimoghaddam et al., 2011). In other words, as a result of salinization, not only in most crops, agricultural production decreases, but also soil physicochemical traits and ecological balance are affected. Salinity disturbs almost every aspect of plant development, including germination process, vegetative growth, and reproduction stages (Ghadirnezhad Shiade et al., 2022b). Shiade & Boelt (2020) studied the germination parameters of nine tall fescue cultivars under different salinity levels and found a delayed germination rate and reduced root and shoot length under higher salinity levels. Similarly, Khare et al. (2020) examined the effects of Na^+, Cl^- and NaCl at a concentration of 100 mM in two rice cultivars and found reduced biomass formation after NaCl > Na^+ > Cl^- in both cultivars.

They attributed the reduced growth of rice plants to additional accumulation of sodium ions leading to lipid peroxidation. Plants experience osmotic stress, ion toxicity, nutrient deficiency (Ca, N, Fe, K, P, Zn), and oxidative stress due to soil salinity (Ghadirnezhad et al., 2020). As phosphate ions co-precipitate with Ca ions in saline soils, phosphorus (P) absorption by plants is significantly reduced (Bano & Fatima, 2009). Plants are particularly sensitive to certain elements such as chlorine, sodium, and boron. Accordingly, osmotic stress and cell death can be rapidly caused by extreme accumulation of sodium in cell walls (Munns, 2002). Depending on the amount of the toxic element in the soil, plants with low salt tolerance may be adversely affected at high salinity levels. On the other hand, high soil salinity may result in imbalanced nutrient levels in plants or affect their ability to take up certain nutrients (Srivastava et al., 2019). The amount of chlorophyll in leaves (Ghadirnezhad Shiade et al., 2022b), stomatal conductance, and photosystem efficiency II are also affected by salinity (Netondo et al., 2004; Shiade & Boelt, 2020), leading to the adverse effect of photosynthetic assimilates supply. Hnilickova et al. (2021) noted that two salinity levels of 100 and 300 mM NaCl reduced stomatal conductance (gs) in the first days of the experiment (day 3), whereas gs, water potential, and maximum quantum yield of the photosystem II (Fv/Fm) decreased with higher salinity levels over time (day 9) in *Portulaca oleracea* L. Reproductive development is affected by salt because it interferes with microsporogenesis and elongation of stamens, promotes abortion of ovules, and causes fertilized

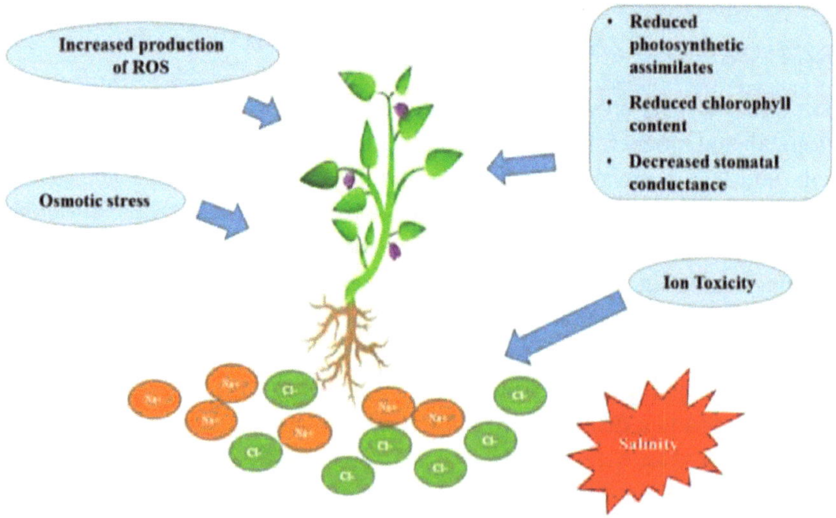

FIGURE 5.1 Various impacts of salinity stress on plants.

embryos to die (Shrivastava & Kumar, 2015). Plants in saline medium require an osmotic balance. When this balance is disturbed, cells lose their turgidity, dehydrate, and eventually die.

The combination of osmotic stress and ion toxicity leads to metabolic imbalance and oxidative stress (Chinnusamy et al., 2010). Ion toxicity can be caused by the replacement of K^+ with Na^+ in biochemical reactions. Several enzymes require K^+ as a cofactor that cannot be replaced by Na^+. In addition to binding tRNA to ribosomes, a high concentration of K^+ is required for protein synthesis (Talaat, 2015). Sodium and chlorine can alter the conformation of proteins. Accordingly, Khare et al. (2020) reported the displacement of K^+ by Na^+ and NaCl, as well as a lower Na^+/K^+ ratio in the more tolerant cultivar (Figure 5.1).

5.3 SALINITY CONSEQUENCE IN THE ENVIRONMENT

Salinization of freshwater bodies, such as streams and rivers, directly influences the quality of drinking and irrigation water through chloride, while sodium and magnesium sulfates affect other organisms (Shrivastava & Kumar, 2015). Habitat fragmentation, habitat loss, and the extinction of intolerant species are among the many consequences of this environmental issue. Salt-induced flocculation can allow harmful algae to grow in water bodies, affecting the ecological health and functions of water ecosystems (Smyth & Elliott, 2016). Having a thinner soil surface layer due to salinity causes soil erosion, which reduces the soil's ability to absorb rainfall, resulting in surface runoff and flooding (Balasubramanian, 2017). Furthermore, both urban and rural communities are affected by salinization, but rural communities are more affected due to high agricultural and commercial losses. Soil permeability and aeration are significantly affected by soil salinization. As a result, saline water accumulates on infrastructure such as concrete structures, brick structures, roads, bridges, pipelines, and buildings, resulting in economic losses and higher costs to repair and maintain the damage (Shrivastava & Kumar, 2015).

5.4 SALINITY STRESS AND NANOTECHNOLOGY-BASED SOLUTIONS

The use of nanotechnology-based solutions is rapidly increasing in various fields of human activity, including agricultural ecosystems, which involve plant food production (Ondrasek et al., 2021). Nanomaterials consist of components that are very small (100 nm or less) and affect the material's traits at the macro level. Compared to larger materials composed of the same mass, nanomaterials have a larger surface area, which changes the chemical properties of the materials and makes them more reactive. In addition, their huge surface area to volume ratio makes them more reactive and potentially more biochemical (Etesami et al., 2021). It has been intensively studied that different forms of metal oxides (ZnO, CuO, Fe_2O_3, Fe_3O_4, CeO_2, TiO_2) can be used as phytonutrients (Kottegoda et al., 2011), plant growth regulators (Gao et al., 2008), and pesticides (Khot et al., 2012). Ghassemi-Golezani & Abdoli (2021) noted that salicylic acid and nano-Fe_2O_3 co-treatment significantly reduced saline toxicity and enhanced Ajowan (Trachyspermum ammi) growth and its essential oil generation. Furthermore, positive effects of Titanium dioxide (TiO_2) nanoparticles (NPs) under salinity were identified in Dracocephalum moldavica (Gohari et al., 2020), and Phaseolus vulgaris (Ondrasek et al., 20220). Moreover, TiO_2 nanoparticles (NPs) were shown to improve wheat seed germination at certain levels (Khodakovskaya et al., 2009). NPs were likely responsible for these effects because of their nanosize, allowing them to penetrate and exert their enhancing functions during growth and treatment (Khodakovskaya et al., 2009). Sarkar & Kalita (2022) confirmed the nanofertilizer and antibacterial activity of Se-NPs to promote the growth of mustard (Brassica nigra) and inhibit phytopathogens under salinity. In the study, the authors suggested that Se-NPs could be used as an antimicrobial agent to treat plant diseases, with little to no toxicity and positive effects on plant growth.

Experiencing a combination of salt stress and nutrient deficiency in semi-arid and arid regions, nanobased applications can have significant benefits due to numerous limitations and relatively high agrochemical application requirements. In addition to their low use efficiency (Ditta & Arshad, 2016), conventional fertilizers can be expensive and technically difficult to apply, and they can remain inactive when in saline (undissolved) form (e.g., in water-stressed environments). Even though fertilizers and treatment agents can be highly mobile, resulting in pollution of special natural resources (e.g., agrochemicals in runoff water) (Ondrasek et al., 2019). In comparison to conventional fertilizer salts, nano-based phytonutrients were proven to be more practical because of their specific functions, reduced losses of nutrients, reduced deterioration of the environment, and boosted use efficiency resulting from gradual and controlled release (Ditta & Arshad, 2016). As an additional benefit, most phytonutrients can be combined into the zeolites nanostructures, a class of minerals composed of silicon and aluminum, which has a large active surface area and far greater cation exchange capacity than soil (Ditta & Arshad, 2016).

5.4.1 NPS TYPES, APPLICATIONS, AND PLANT UPTAKE

The field of nanotechnology includes the evaluation and application of materials with a small size of 1–100 nm on at least one side (width, length, and height) that possess a unique property (Etesami et al., 2021). Because nanomaterials are smaller and have a larger surface area, they differ from bulk materials in terms of electrical conductivity, chemical reactivity, magnetism, optical effects, and physical strength (Asha & Narain, 2020). A nanomaterial can be either one-dimensional (e.g., surface films), two-dimensional (such as fibers or strands), or three-dimensional (such as particles). The shapes can be tubular, spherical, or irregular and the type can be single, fused, aggregated, or agglomerated (Das & Das, 2019). Furthermore, according to their chemical traits, NPs can be categorized into four key categories: (a) metal nanoparticles such as Ag, Au, Pt, Zn, and Ni; (b) carbon nanoparticles including fullerenes, graphene, single-walled carbon nanotubes, and so on; (c) polymer compounds; and (d) metal oxides (ceramics) such as ZnO, TiO_2, and FeO_2 (Etesami et al., 2021). Table 5.1 presents different NPs used in experiments to relieve salinity.

Since NPs are physicochemically different, their concentrations and application methods determine their biological function (Etesami et al., 2021). Various methods are used to apply NPs, such as irrigation, priming, hydroponic compounds, foliar spraying, and direct injection (Abou-Zeid & Ismail, 2018). Their mechanism of action is to penetrate plant tissues via root junctions and wound areas and infiltrate the cell wall and membrane tissues of the root epidermis (through various mechanisms, including endocytosis, plasmodesmata, carrier proteins, and pores). A whole series of events can accompany this phenomenon to reach the xylem (plant vascular bundle) (via the apoplast or symplast) (Pérez-de-Luque, 2017), collect in cells or subcellular organelles, and then transfer simply to the stele, resulting in translocation in the leaf (Pérez-de-Luque, 2017). In addition, nanoparticles can penetrate the cuticle, hydathodes, stomata, and trichrome of leaves and reach the cell cytoplasm (Etesami et al., 2021). Because NPs bind to various organelles in the cytoplasm, they can affect metabolic processes at these sites (Etesami et al., 2021). NPs can also be taken up directly from seeds via parenchymal intercellular tissues; afterwards, the diffusion occurs in cotyledon (Etesami et al., 2021).

TABLE 5.1 Different NPs Types Used to Enhance Plant Tolerance to Salinity

Nanobiochar type	Crop	Effect	References
Multi-walled C nanotubes (10–40 mg L^{-1})	Tomato (*Solanum lycopersicum* L)	Enhanced the seed germination. Improved plant growth.	Khodakovskaya et al. (2009)
Multi-walled C nanotubes and graphene NPs	switch grass (*Panicum virgatum*)	Improved seed germination rate. Boosted plant total biomass.	Pandey et al. (2018)
ZnO nanoparticles	Coffee (*Coffea arabica)*	Positively impacted plant growth and physiology yield.	Rossi et al. (2019)
Fe_2O_3 nanoparticles	Peanut (*Arachis hypogaea*)	Exhibited effective impacts to be replaced by Fe traditional fertilizers	Rui et al. (2016)
Ag nanoparticles	Wheat (*Triticum aestivum*)	Boosted plant growth Enhanced proline and soluble sugars contents. Improved POD activity,whereas decreased CAT activity.	Mohamed et al. (2017)
SiO_2 nanoparticles	Plants	Improve plant water content	Zulfiqar & Ashraf (2021)
Si nanoparticles	Wheat (*Triticum aestivum*)	Improved chlrophyl content. Enhanced germination.	Mushtaq et al. (2019)
Cu nanoparticles	Tomato (*Lycopersicon esculentum*)	Improved antioxidant activity, phenols, and glutathione. Regulated Na+/K+ ratio.	Pérez-Labrada et al. (2019)
Ce nanoparticles	Canola (*Brassica napus*)	Improved photosynthetic efficacy.	Rossi et al. (2016)
Ce nanoparticles	Canola (*Brassica napus*)	Higher Na+ transport to areal organs, causing less Na+ accumulation of in roots	Rossi et al. (2017)

5.4.2 SALINITY STRESS AND THE EFFECTS OF NPS ON PLANTS

There is evidence that different NPs promote the growth and development of salt-stressed plants at concentrations below certain limits through various mechanisms (Etesami et al., 2021) (Figure 5.2).

At the molecular level, there is evidence that NPs can influence gene expression to alleviate salt stress. For example, the expression of microRNAs was examined under treatment with NPs and found increased expression

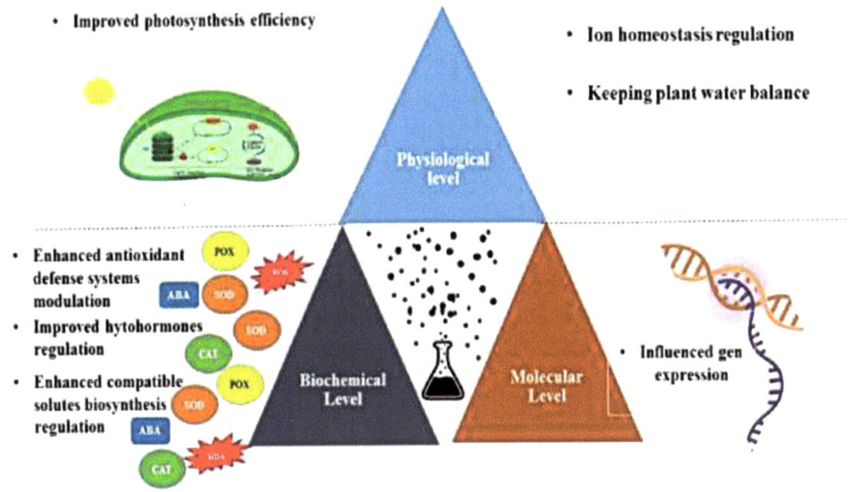

FIGURE 5.2 Various physiological, biochemical, and biochemical functions of nanobiochar to enhance plant tolerance under salinity stress. POX: peroxidase; SOD: superoxide dismutase; CAT: catalase; ABA: Abscisic acid; MDA: malondialdehyde; ROS: reactive oxygen species.

of miR408 and miR398, which regulate seed germination, seedling shoot and root growth, and antioxidant scavenging mechanisms (Kumar et al., 2013). This enhancement prevented the aforementioned factors. In another study, Zn-NPs were applied foliarly to rapeseed under salt stress, resulting in changes in gene expression upon stress. The expression of SKRD2, MYC, and MPK4 genes was reduced, while others (e.g., MPK and ARP) were increased. There are a number of physiological, hormonal, and growth reactions associated with SKRD2 and MYC genes. These are transcription factors improving abiotic stress tolerance (Etesami et al., 2021).

In terms of biochemical aspects, three action mechanisms can be mentioned, such as modulation of antioxidant defense systems, regulation of phytohormones, and biosynthesis of compatible solutes (Etesami et al., 2021). The generation of ROS in plants is a reaction to abiotic stress, including salinity. To cope with the excessive production of ROS in salt-stressed cells, plants have evolved antioxidant enzymes (Ghadirnezhad et al., 2020). Numerous studies have confirmed that NPs increase the levels of antioxidant enzymes (Abdoli et al., 2020; Etesami et al., 2021). Farouk & Al-Amri (2019) identified that exogenous application of Zinc (Zn)-NPs counteracted NaCl-induced impairment via regulation of the antioxidant system, osmotic and ion adjustment in Canola (Brassica napus L. cv. Pactol). As a result of increased salinity, membrane permeability (MP %), hydrogen peroxide (H_2O_2), and malondialdehyde (MDA) contents increased. Zinc foliar spraying, on the other hand, significantly decreased MP %, MDA, and H_2O_2 concentrations, which can be attributed to increased antioxidant enzymes (superoxide dismutase (SOD), catalase (CAT), peroxidase (POX)) and nonenzymatic antioxidants (ascorbic acids, carotenoids, and total soluble phenolic compounds). Additionally, zinc application increases total soluble carbohydrates and proline accumulation, potassium, nitrogen, and phosphorus content in plant tissues, while sodium and chloride content declines. Using similar methods, Ti-NPs significantly mitigated the adverse effects of salinity on marigolds (Calendula officinalis L.). The authors studied the Ti-NPs foliar spraying at three levels (0, 50, 100, and 200 mg L^{-1}) under diverse levels of salinity (0, 30, 60, and 90 mM). They found that TiO_2 NPs at 100 mg L^{-1} enhanced the leaves' flavonoid content, CAT, and SOD activities, whereas decreased MDA and H_2O_2 contents. In another study, Wahid et al. (2020) highlighted the effect of Silver (Ag)-NPs in reducing ABA concentration (approximately 41%) in wheat plants exposed to salinity stress. The ABA-mediated phytohormone pathway has been explored in the context of salt-induced stress because it provides critical physiological responses against stressful environmental conditions (Wahid et al., 2020). Plants under stress conditions employ a variety of strategies to reduce the diffusion of Na^+ ions from the thick caspase strips into the endodermis, resulting in a positive correlation between the biosynthesis of ABA and traits of salinity tolerance, such as inducing seed dormancy, regulating turgor pressure, and reducing water loss by partially or completely closing stomatal apertures (Sah et al., 2016).

As for physiological effects, it is assumed that NPs can influence plant growth and development via three mechanisms of photosynthesis, ion homeostasis regulation, and maintaining plant water balance (Etesami et al., 2021). According to Lashkary et al. (2021), TiO_2NPs (at a concentration of 200 mg L^{-1}) enhanced photosynthetic pigments content, carotenoid content, and total soluble sugar in marigold under salinity. Additionally, Wahid et al. (2020) documented that AgNPs increased stomatal area by approximately 10%, chlorophyll concentration, and Plant Dry Mass. The authors argued that enhanced chlorophyll with AgNPs leads to improved PS-II quantum efficiency and increased fluorescence parameters, resulting in higher photosynthetic assimilation production (Wahid et al., 2020). As a result of the effect of AgNPs on the stomatal surface, turgor pressure in guard cells is altered, aperture acclimation is enhanced, and gas exchange rates are optimized in response to abiotic stresses (Wahid et al., 2020).

Several studies have shown that high potassium-to-sodium ratios, which are disturbed by salinity, are important for plant resistance to salt stress. Sytar et al. (2019) have demonstrated that NPs can increase this ratio, thereby improving the plant's osmotic potential and helping it to survive salinity stress. As an example, Farhangi-Abriz & Torabian (2018) found that Nano SiO_2 improved soybean plantlet growth under stress by boosting leaf K^+ concentrations. Ye et al. (2020) investigated the effects of Mn-NPs on pepper exposed to salinity and found that Mn-NPs increased root growth and reduced the Na, Ca, Mn, and K redistribution between roots and aerial organs in comparison with control salt-exposed plants. The amount and rate of water absorption by plants under salinity stress decrease. By increasing osmotic stress in soil solutions, salts limit the absorption of water by plants, impacting leaf water content, stomatal conductance (g_s), leaf growth (accelerating leaf senescence and causing leaf death), photosynthesis (reduction in chlorophyll concentrations), and eventually reducing plant growth (Munns, 2005). The water balance of plants is controlled by aquaporins, a type of channel protein that belongs to the family of large intrinsic proteins and assists in the transport of water and other small neutral molecules across cell membranes (Kapilan et al., 2018). The improved expression of plasma-membrane intrinsic proteins, aquaporins, by NPs has been found to increase root hydraulic conductance, which may also contribute to a reduction in oxidative stress, membrane damage, and increased water uptake (Etesami et al., 2021).

5.5 SALINITY AND NANO-BIOCHAR

Biochar is produced by heating biomass feedstocks such as agricultural waste and lignocellulosic resources to a controlled decomposition temperature without oxygen or with limited oxygen (pyrolysis). Different thermochemical processes can also produce biochar from solid wastes and animal litter. Scientists are interested in biochar technology because it can mitigate climate change, mitigate abiotic stressors, and boost plant growth, soil health, and fertility (Chausali et al., 2021). The production of biochar is not only economical and environmentally friendly, but can also be very efficient in reusing waste resources because it requires low energy and can be produced at temperatures as high as 700°C (Chausali et al., 2021). By reducing the particle size in the micro range (10–600 nm), more adsorption sites are available, resulting in better adsorption capacity (Lonappan et al.,

2016). Furthermore, nanosized biochar particles that are at least 100 nm or smaller have improved properties due to the larger surface area to volume ratio, resulting in higher surface energy, adsorption potential, and biological efficacy (Naghdi et al., 2019).

5.5.1 NANOBOCHAR PROPERTIES

Low-temperature nanobiochars have surface areas of 5.6 to 47.2 m^2 g^{-1}, respectively, whereas high-temperature nanobiochars have significantly higher surface areas (342 to 430 m^2 g^{-1}) at 450 and 600C (Ramanayaka et al., 2020). In comparison to biochar at the same temperature, rice straw nanobiochar at 400°C displayed a notable decrease in surface area from 141.3 to 93.28 m^2/g. In contrast, macrobiochar had a smaller surface area at 700°C than nanobiochar. The surface area of rice straw nanobiochar increased when the temperature was increased to 700°C, as expected (Lian et al., 2018). In terms of water-holding capacity, biochar's hydrophilic traits are directly influenced by the pyrolysis process and parent material. A higher pyrolysis temperature reduces the number of functional groups, leading to a decrease in water binding capacity and an increase in hydrophobicity (Ramanayaka et al., 2020). Accordingly, a greater number of functional groups enhances affinity towards water absorption (Weber & Quicker, 2018). The water-holding capacities of pinewood nanobiochars obtained by pyrolysis at 525C and 400°C were reported to be 9.75 and 2.7, respectively (Naghdi et al., 2019). There is a negative correlation between mechanical stability and the porosity of biochar. The low strength of highly porous biochar results from processing methods such as grinding, which cause cracks in the structure, leading to volatile matter leakage and water evaporation. As a result, nanobiochar is generally low in strength (Ramanayaka et al., 2020).

5.5.2 BIOCHAR-BASED METAL OXIDE NANOCOMPOSITES (BNC)

New categories of biochar materials, such as biochar-based metal oxide nanocomposites (BNC), combine biochar's benefits with nanomaterials (Liu et al., 2015). Changing the composition of biochar modifies its physicochemical properties, which increases carbon sequestration and soil fertility (Ghassemi-Golezani & Farhangi-Abriz, 2021). Physicochemical characteristics of these materials, such as their high specific surface area and cation exchange capacity, cause them to provide high fertilization capacity in agriculture and sufficient nutrients compounds in the soil (Ghassemi-Golezani & Farhangi-Abriz, 2021). The same authors evaluated the potential effects of BNC at various concentrations on phenological, morphological, and yield traits of the safflower crop (Carthamus tinctorius L.) under salinity. They found that biochar and BNC-MgO combination, particularly BNC-MgO treatment, enhanced the cation exchange capacity (CEC) and pH of the soil, the main and lateral roots extension and density, chlorophyll stability and content, branches, and flowers number per plant of safflower in saline conditions. The positive effect of BNC-MgO to improve CEC can be explained by its ability to mend organic matter content and decrease exchangeable sodium percentage and its uptake in saline soil (Ghassemi-Golezani & Farhangi-Abriz, 2021). Furthermore, BNC-MgO noticeably absorbs sodium ions, leading to a reduced availability of this ion in the plants' organs in stressful situations. In the rhizosphere, nanocomposite materials have a high specific surface area, which can increase the active surface area for cation exchange. The high CEC (cation exchange capacity) ability of BNCs enables them to exchange nutrients in the rhizosphere. Additionally, the higher CEC capability of saline soils under BNC treatment can be explained by their potential to regulate soil pH (Ghassemi-Golezani & Farhangi-Abriz, 2021). Since BNCs mainly contribute to the uptake of nutrients, they are more available under salinity, which promotes root growth and expansion.

5.6 CONCLUSION AND FUTURE PROSPECTS

There are many agricultural areas around the world that are negatively affected by salinity stress, one of the most damaging non-biological stresses. Plants are affected by salt stress in many ways, including morphological, physiological, and biochemical changes. This environmental challenge is exacerbated by climate change. The use of nanotechnology is rapidly increasing in various fields, especially in agriculture, as it offers potential and rapid solutions that are essential for achieving sustainability. In new classes of biochar materials, such as nanobiochar, the advantages of biochar are combined with nanomaterials that have improved physicochemical properties. These materials supply high fertilizing capacity in agriculture and provide adequate amounts of nutrients due to their high specific surface area and cation exchange capacity, particularly under salinity. To elucidate how nanobiochar triggers specific metabolic processes in salt-stressed plants, further research in this area is recommended. A deeper understanding of the potential of these compounds in alleviating salt stress, as well as assessing their risks to plants and the environment, will make them more valuable.

KEYWORDS

- **biochar**
- **environment**
- **nanobiochar**
- **nanotechnology**
- **salinity**

REFERENCES

Abdoli, S., Ghassemi-Golezani, K., & Alizadeh-Salteh, S., (2020). Responses of ajowan (*Trachyspermum ammi* L.) to exogenous salicylic acid and iron oxide nanoparticles under salt stress. *Environ. Sci. Pollut. Res., 27*, 36939–36953.

Abeer, H., Abd_Allah, E. F., Alqarawi, A. A., El-Didamony, G., Alwhibi, M., Egamberdieva, D., & Ahmad, P., (2014). Alleviation of adverse impact of salinity on faba bean (*Vicia faba* L.) by arbuscular mycorrhizal fungi. *Pak. J. Bot., 46*, 2003–2013.

Abou-Zeid, H., & Ismail, G., (2018). The role of priming with biosynthesized silver nanoparticles in the response of *Triticum aestivum* L to salt stress. *Egypt. J. Bot., 58*, 73–85.

Akbarimoghaddam, H., Galavi, M., Ghanbari, A., & Panjehkeh, N., (2011). Salinity effects on seed germination and seedling growth of bread wheat cultivars. *Trakia J. Sci., 9*, 43–50.

Ali, S., Rizwan, M., Qayyum, M. F., Ok, Y. S., Ibrahim, M., Riaz, M., Arif, M. S., et al., (2017). Biochar soil amendment on alleviation of drought and salt stress in plants: A critical review. *Environ. Sci. Pollut. Res., 24*, 12700–12712. https://doi.org/10.1007/s11356-017-8904-x.

Asha, A. B., & Narain, R., (2020). Nanomaterials properties. In: *Polymer Science and Nanotechnology* (pp. 343–359). Elsevier.

Balasubramanian, A., (2017). *Soil Erosion–Causes and Effects.* Cent. Adv. Stud. Earth Sci. Univ. Mysore, Mysore. https://doi.org/10.13140/RG.2.2.26247.39841.

Bano, A., & Fatima, M., (2009). Salt tolerance in *Zea mays* (L). following inoculation with *Rhizobium* and *Pseudomonas. Biol. Fertil. Soils, 45*, 405–413.

Chaitanya, K. V., Krishna, C. R., Ramana, G. V., & Beebi, S. K., (2014). Salinity stress and sustainable agriculture: A review. *Agric. Rev., 35*, 34–41.

Chausali, N., Saxena, J., & Prasad, R., (2021). Nanobiochar and biochar based nanocomposites: Advances and applications. *J. Agric. Food Res., 5*, 100191. https://doi.org/10.1016/j.jafr.2021.100191.

Chen, M., Wang, D., Yang, F., Xu, X., Xu, N., & Cao, X., (2017). Transport and retention of biochar nanoparticles in a paddy soil under environmentally-relevant solution chemistry conditions. *Environ. Pollut., 230*, 540–549.

Chinnusamy, V., Zhu, J. K., & Sunkar, R., (2010). Gene regulation during cold stress acclimation in plants. *Plant Stress Toler. Methods Protoc.*, 39–55.

Das, A., & Das, B., (2019). Nanotechnology a potential tool to mitigate abiotic stress in crop plants. *Abiotic Biot. Stress Plants.*

Ditta, A., & Arshad, M., (2016). Applications and perspectives of using nanomaterials for sustainable plant nutrition. *Nanotechnol. Rev.*, *5*, 209–229.

Etesami, H., Fatemi, H., & Rizwan, M., (2021a). Interactions of nanoparticles and salinity stress at physiological, biochemical and molecular levels in plants: A review. *Ecotoxicol. Environ. Saf.*, *225*. https://doi.org/10.1016/j.ecoenv.2021.112769.

Farhangi-Abriz, S., & Ghassemi-Golezani, K., (2021). Changes in soil properties and salt tolerance of safflower in response to biochar-based metal oxide nanocomposites of magnesium and manganese. *Ecotoxicol. Environ. Saf.*, *211*, 111904. https://doi.org/10.1016/j.ecoenv.2021.111904.

Farhangi-Abriz, S., & Torabian, S., (2018). Nano-silicon alters antioxidant activities of soybean seedlings under salt toxicity. *Protoplasma*, *255*, 953–962.

Farouk, S., & Al-Amri, S. M., (2019). Exogenous zinc forms counteract NaCl-induced damage by regulating the antioxidant system, osmotic adjustment substances, and ions in canola (*Brassica napus* L. cv. Pactol) plants. *J. Soil Sci. Plant Nutr.*, *19*, 887–899.

Gao, F., Liu, C., Qu, C., Zheng, L., Yang, F., Su, M., & Hong, F., (2008). Was improvement of spinach growth by nano-TiO_2 treatment related to the changes of *Rubisco activase*? *Biometals*, *21*, 211–217.

Ghadirnezhad, S. R., Esmaeili, M., Hematollah, P., & Nematzade, G., (2020). Physiological and biochemical evaluation of sixth generation of rice (*Oryza sativa* L.) mutant lines under salinity stress. *J. Plant Process Funct.*, *9*, 57–72.

Ghadirnezhad, S. S. R., Fathi, A., Taghavi, G. F., Amiri, E., & Pessarakli, M., (2023). Plants' responses under drought stress conditions: Effects of strategic management approaches – A review. *Journal of Plant Nutrition*, *46*(9), 2198–2230.

Ghadirnezhad, S. S. R., Pirdashti, H., Esmaeili, M. A., & Nematzade, G. A., (2022b). Biochemical and physiological characteristics of mutant genotypes in rice (*Oryza sativa* L.) contributing to salinity tolerance indices. *Gesunde Pflanz.* https://doi.org/10.1007/s10343-022-00701-7.

Ghassemi-Golezani, K., & Abdoli, S., (2021). Improving ATPase and PPase activities, nutrient uptake and growth of salt stressed ajowan plants by salicylic acid and iron-oxide nanoparticles. *Plant Cell Rep.*, *40*, 559–573. https://doi.org/10.1007/s00299-020-02652-7.

Ghassemi-Golezani, K., & Farhangi-Abriz, S., (2021). Biochar-based metal oxide nanocomposites of magnesium and manganese improved root development and productivity of safflower (*Carthamus tinctorius* L.) under salt stress. *Rhizosphere*, *19*. https://doi.org/10.1016/j.rhisph.2021.100416.

Gohari, G., Mohammadi, A., Akbari, A., Panahirad, S., Dadpour, M. R., Fotopoulos, V., & Kimura, S., (2020). Titanium dioxide nanoparticles (TiO_2NPs) promote growth and ameliorate salinity stress effects on essential oil profile and biochemical attributes of *Dracocephalum moldavica*. *Sci. Rep.*, *10*, 1–14. https://doi.org/10.1038/s41598-020-57794-1.

Hnilickova, H., Kraus, K., Vachova, P., & Hnilicka, F., (2021). Salinity stress affects photosynthesis, malondialdehyde formation, and proline content in *Portulaca oleracea* L. *Plants*, *10*, 845.

Hofmann, T., Lowry, G. V., Ghoshal, S., Tufenkji, N., Brambilla, D., Dutcher, J. R., Gilbertson, L. M., et al., (2020). Technology readiness and overcoming barriers to sustainably implement nanotechnology-enabled plant agriculture. *Nat. Food, 1*, 416–425.

Imadi, S. R., Shah, S. W., Kazi, A. G., Azooz, M. M., & Ahmad, P., (2016). Phytoremediation of saline soils for sustainable agricultural productivity. *Plant Met. Interact.*, 455–468.

Kapilan, R., Vaziri, M., & Zwiazek, J. J., (2018). Regulation of aquaporins in plants under stress. *Biol. Res., 51*, 1–11.

Khare, T., Srivastava, A. K., Suprasanna, P., & Kumar, V., (2020). Individual and additive stress impacts of Na+ and Cl^- on proline metabolism and nitrosative responses in rice. *Plant Physiol. Biochem., 152*, 44–52. https://doi.org/https://doi.org/10.1016/j.plaphy.2020.04.028.

Khodakovskaya, M., Dervishi, E., Mahmood, M., Xu, Y., Li, Z., Watanabe, F., & Biris, A. S., (2009). Carbon nanotubes are able to penetrate plant seed coat and dramatically affect seed germination and plant growth. *ACS Nano, 3*, 3221–3227.

Khot, L. R., Sankaran, S., Maja, J. M., Ehsani, R., & Schuster, E. W., (2012). Applications of nanomaterials in agricultural production and crop protection: A review. *Crop Prot., 35*, 64–70.

Kottegoda, N., Munaweera, I., Madusanka, N., & Karunaratne, V., (2011). A green slow-release fertilizer composition based on urea-modified hydroxyapatite nanoparticles encapsulated wood. *Curr. Sci.*, 73–78.

Kumar, V., Guleria, P., Kumar, V., & Yadav, S. K., (2013). Gold nanoparticle exposure induces growth and yield enhancement in *Arabidopsis thaliana. Sci. Total Environ., 461*, 462–468.

Lashkary, M., Moghaddam, M., Asgharzade, A., & Tatari, M., (2021). Titanium dioxide nanoparticle is involved in mitigating NaCl-induced *Calendula officinalis* L. by activation of antioxidant defense system and accumulation of osmolytes. *Plant Physiol. Biochem., 166*, 31–40.

Lian, F., Yu, W., Wang, Z., & Xing, B., (2018). New insights into black carbon nanoparticle-induced dispersibility of goethite colloids and configuration-dependent sorption for phenanthrene. *Environ. Sci. Technol., 53*, 661–670.

Liang, W., Ma, X., Wan, P., & Liu, L., (2018). Plant salt-tolerance mechanism: A review. *Biochem. Biophys. Res. Commun., 495*, 286–291.

Liu, P., Yin, L., Wang, S., Zhang, M., Deng, X., Zhang, S., & Tanaka, K., (2015). Enhanced root hydraulic conductance by aquaporin regulation accounts for silicon alleviated salt-induced osmotic stress in *Sorghum bicolor* L. *Environ. Exp. Bot., 111*, 42–51.

Lonappan, L., Rouissi, T., Das, R. K., Brar, S. K., Ramirez, A. A., Verma, M., Surampalli, R. Y., & Valero, J. R., (2016). Adsorption of methylene blue on biochar microparticles derived from different waste materials. *Waste Manag., 49*, 537–544. https://doi.org/https://doi.org/10.1016/j.wasman.2016.01.015.

Mahmoud, E., El Baroudy, A., Ali, N., & Sleem, M., (2020). Spectroscopic studies on the phosphorus adsorption in salt-affected soils with or without nano-biochar additions. *Environ. Res., 184*, 109277. https://doi.org/10.1016/j.envres.2020.109277.

Mavi, M. S., Marschner, P., Chittleborough, D. J., Cox, J. W., & Sanderman, J., (2012). Salinity and sodicity affect soil respiration and dissolved organic matter dynamics differentially in soils varying in texture. *Soil Biol. Biochem., 45*, 8–13.

Mindari, W., Sasongko, P. E., Kusuma, Z., Syekhfani, & Aini, N., (2018). Efficiency of various sources and doses of humic acid on physical and chemical properties of saline soil and growth and yield of rice. In: *AIP Conference Proceedings* (p. 30001). AIP Publishing LLC. https://doi.org/10.1063/1.5061854.

Mohamed, A. K. S. H., Qayyum, M. F., Abdel-Hadi, A. M., Rehman, R. A., Ali, S., & Rizwan, M., (2017). Interactive effect of salinity and silver nanoparticles on photosynthetic and biochemical parameters of wheat. *Arch. Agron. Soil Sci., 63*, 1736–1747.

Munns, R., (2002). Comparative physiology of salt and water stress. *Plant. Cell Environ., 25*, 239–250.

Munns, R., (2005). Genes and salt tolerance: Bringing them together. *New Phytol., 167*, 645–663.

Mushtaq, A., Rizwan, S., Jamil, N., Ishtiaq, T., Irfan, S., Ismail, T., Malghani, M. N., & Shahwani, M. N., (2019). Influence of silicon sources and controlled release fertilizer on the growth of wheat cultivars of balochistan under salt stress. *Pak. J. Bot., 51*, 1561–1567.

Naghdi, M., Taheran, M., Pulicharla, R., Rouissi, T., Brar, S. K., Verma, M., & Surampalli, R. Y., (2019). Pine-wood derived nanobiochar for removal of carbamazepine from aqueous media: Adsorption behavior and influential parameters. *Arab. J. Chem., 12*, 5292–5301. https://doi.org/https://doi.org/10.1016/j.arabjc.2016.12.025.

Negacz, K., Malek, Ž., De Vos, A., & Vellinga, P., (2022). Saline soils worldwide: Identifying the most promising areas for saline agriculture. *J. Arid Environ., 203*. https://doi.org/10.1016/j.jaridenv.2022.104775.

Netondo, G. W., Onyango, J. C., & Beck, E., (2004). Sorghum and salinity: II. Gas exchange and chlorophyll fluorescence of sorghum under salt stress. *Crop Sci., 44*, 806–811.

Ondrasek, G., Begić, H. B., Zovko, M., Filipović, L., Meriño-Gergichevich, C., Savić, R., & Rengel, Z., (2019). Biogeochemistry of soil organic matter in agroecosystems & environmental implications. *Sci. Total Environ., 658*, 1559–1573.

Ondrasek, G., Kovačić, M. B., Carević, I., Štirmer, N., Stipičević, S., Udiković-Kolić, N., Filipović, V., et al., (2021). Bioashes and their potential for reuse to sustain ecosystem services and underpin circular economy. *Renew. Sustain. Energy Rev., 151*, 111540.

Ondrasek, G., Rathod, S., Manohara, K. K., Gireesh, C., Anantha, M. S., Sakhare, A. S., Parmar, B., et al., (2022). Salt stress in plants and mitigation approaches. *Plants, 11*, 1–21.

Pandey, K., Lahiani, M. H., Hicks, V. K., Hudson, M. K., Green, M. J., & Khodakovskaya, M., (2018). Effects of carbon-based nanomaterials on seed germination, biomass accumulation and salt stress response of bioenergy crops. *PLoS One, 13*, e0202274.

Parihar, P., Singh, S., Singh, R., Singh, V. P., & Prasad, S. M., (2015). Effect of salinity stress on plants and its tolerance strategies: A review. *Environ. Sci. Pollut. Res., 22*, 4056–4075. https://doi.org/10.1007/s11356-014-3739-1.

Patel, B. B., & Dave, R. S., (2011). Studies on infiltration of saline-alkali soils of several parts of Mehsana and Patan districts of North Gujarat. *J. Appl. Technol. Environ. Sanit., 1*, 87–92.

Pérez-de-Luque, A., (2017). Interaction of nanomaterials with plants: What do we need for real applications in agriculture? *Front. Environ. Sci., 5*, 12.

Pérez-Labrada, F., López-Vargas, E. R., Ortega-Ortiz, H., Cadenas-Pliego, G., Benavides-Mendoza, A., & Juárez-Maldonado, A., (2019). Responses of tomato plants under saline stress to foliar application of copper nanoparticles. *Plants, 8*, 151.

Rajput, V. D., Minkina, T., Ahmed, B., Singh, V. K., Mandzhieva, S., Sushkova, S., Bauer, T., et al., (2022). Nano-biochar: A novel solution for sustainable agriculture and environmental remediation. *Environ. Res., 210*, 112891. https://doi.org/10.1016/j.envres.2022.112891.

Ramanayaka, S., Vithanage, M., Alessi, D. S., Liu, W. J., Jayasundera, A. C. A., & Ok, Y. S., (2020). Nanobiochar: Production, properties, and multifunctional applications. *Environ. Sci. Nano, 7*, 3279–3302. https://doi.org/10.1039/d0en00486c.

Rossi, L., Fedenia, L. N., Sharifan, H., Ma, X., & Lombardini, L., (2019). Effects of foliar application of zinc sulfate and zinc nanoparticles in coffee (*Coffea arabica* L.) plants. *Plant Physiol. Biochem., 135*, 160–166.

Rossi, L., Zhang, W., & Ma, X., (2017). Cerium oxide nanoparticles alter the salt stress tolerance of *Brassica napus* L. by modifying the formation of root apoplastic barriers. *Environ. Pollut., 229*, 132–138.

Rossi, L., Zhang, W., Lombardini, L., & Ma, X., (2016). The impact of cerium oxide nanoparticles on the salt stress responses of *Brassica napus* L. *Environ. Pollut., 219*, 28–36.

Rui, M., Ma, C., Hao, Y., Guo, J., Rui, Y., Tang, X., Zhao, Q., et al., (2016). Iron oxide nanoparticles as a potential iron fertilizer for peanut (*Arachis hypogaea*). *Front. Plant Sci., 7*, 815.

Ruto, E., Gould, I., Bosworth, G., & Wright, I., (2018). *The State, Causes and Impact of Soil Salinization: A Global Overview.* SALFAR work packag. 3. Baseline January 2018 Report, Lincoln.

Sah, S. K., Reddy, K. R., & Li, J., (2016). Abscisic acid and abiotic stress tolerance in crop plants. *Front. Plant Sci., 7*, 571.

Sarkar, R. D., & Kalita, M. C., (2022). Se nanoparticles stabilized with *Allamanda cathartica* L. flower extract inhibited phytopathogens and promoted mustard growth under salt stress. *Heliyon, 8*, e09076.

Shiade, S. R. G., & Boelt, B., (2020). Seed germination and seedling growth parameters in nine tall fescue varieties under salinity stress. *Acta Agric. Scand. Sect. B Soil Plant Sci., 70*, 485–494. https://doi.org/10.1080/09064710.2020.1779338.

Shrivastava, P., & Kumar, R., (2015). Soil salinity: A serious environmental issue and plant growth promoting bacteria as one of the tools for its alleviation. *Saudi J. Biol. Sci., 22*, 123–131.

Smyth, K., & Elliott, M., (2016). Effects of changing salinity on the ecology of the marine environment. In: Solan, M., & Whiteley, N., (eds.), *Stressors in the Marine Environment; Physiological and Ecological Responses, Societal Implications* (pp. 161–174). Oxford University Press, Oxford. https://doi.org/10.1093/acprof:oso/9780198718826.003.0009.

Srivastava, P., Wu, Q., & Giri, B., (2019). Salinity: An overview, In: Giri, B., & Varma, A., (eds.), *Microorganisms in Saline Environments: Strategies and Functions* (pp. 3–18). Springer Nature Switzerland AG. https://doi.org/10.1007/978-3-030-18975-4_1.

Sytar, O., Kumari, P., Yadav, S., Brestic, M., & Rastogi, A., (2019). Phytohormone priming: Regulator for heavy metal stress in plants. *J. Plant Growth Regul., 38*, 739–752.

Talaat, N. B., (2015). Effective microorganisms modify protein and polyamine pools in common bean (*Phaseolus vulgaris* L.) plants grown under saline conditions. *Sci. Hortic., 190*, 1–10.

Wahid, I., Kumari, S., Ahmad, R., Hussain, S. J., Alamri, S., Siddiqui, M. H., & Khan, M. I. R., (2020). Silver nanoparticle regulates salt tolerance in wheat through changes in ABA concentration, ion homeostasis, and defense systems. *Biomolecules, 10*, 1506.

Wang, P., Lombi, E., Zhao, F. J., & Kopittke, P. M., (2016). Nanotechnology: A new opportunity in plant sciences. *Trends Plant Sci., 21*, 699–712.

Weber, K., & Quicker, P., (2018). Properties of biochar. *Fuel, 217*, 240–261.

Ye, Y., Cota-Ruiz, K., Hernandez-Viezcas, J. A., Valdes, C., Medina-Velo, I. A., Turley, R. S., Peralta-Videa, J. R., & Gardea-Torresdey, J. L., (2020). Manganese nanoparticles control salinity-modulated molecular responses in *Capsicum annuum* L. through priming: A sustainable approach for agriculture. *ACS Sustain. Chem. Eng., 8*, 1427–1436.

Zhao, G., Zhao, Y., Lou, W., Su, J., Wei, S., Yang, X., Wang, R., et al., (2019). Nitrate reductase-dependent nitric oxide is crucial for multi-walled carbon nanotube-induced plant tolerance against salinity. *Nanoscale, 11*, 10511–10523.

Zulfiqar, F., & Ashraf, M., (2021a). Plant physiology and biochemistry nanoparticles potentially mediate salt stress tolerance in plants. *Plant Physiol. Biochem., 160*, 257–268. https://doi.org/https://doi.org/10.1016/j.plaphy.2021.01.028.

CHAPTER 6

Enhancing the Nutrient Use Efficiency Through Nano-Biochar

PARAMITA DEB

School of Agricultural Sciences, GD Goenka University, Sohna, Haryana, India

ABSTRACT

The world's population is also growing every day. The level of nourishment in the form of nutrients in the earth has significantly decreased because of industrialization, deforestation, overgrazing, and other human activities. Food security suffers significantly due to decreased production and plant development caused by nutrient deficiency. Chemical fertilizers have been overused to preserve soil nutrients. However, this has led to nitrogen loss and runoff, reducing their bioavailability in the soil and promoting the use of chemical-based fertilizers. Feeding such a large population will undoubtedly have a very negative impact on the soil and biosphere in order to increase crop output. Farmers rely exclusively on chemicals to improve agricultural production and harvests, which degrade the environment and pose health risks to people. Nano biochar has better chemical and physical properties. It can have various elemental compositions, such as a polar or aromatic nature, pH, high cation exchange capacity (CEC), larger specific surface area (SSA), pore properties, and Zeta potential. It has many uses, including improving soil quality and plant growth, controlling plant diseases, bioremediating pesticides and contaminants, treating wastewater, and acting as a support material for enzyme immobilization. It may prove to be the best alternative to conventional methods due to its affordability, sustainability, and environmental friendliness. Nano-biochar is a viable substitute for chemical fertilizers because it is mobile in soil and excels at absorbing pollutants, nutrients, and toxins.

Nanomaterials and Nano-Biochar in Reducing Soil Stress: An Integrated Approach to Sustainable Agriculture. Vishnu D. Rajput, & Tatiana Minkina (Eds.)

6.1 INTRODUCTION

The population of the world has increased fivefold in the previous century (from 1.6 billion in 1900 to 8.0 billion in 2022), and the agricultural industry has been under extreme strain to keep up with the rapid rise in food consumption. According to a recent projection, the world's food supply will need to expand by 110 percent by the year 2050 in order to meet societal food demand (Tilman et al., 2011). According to scientific estimates, agricultural production should increase by 1.4% annually (OECD, 2018), but others predicted that by 2050, food demand will need to double, and 60% more cereal production will be needed to meet the necessary levels of demand (FAO, 2009; Glenn et al., 2008; Baulcombe et al., 2009; Tilman et al., 2011). A major improvement in water use efficiency (WUE) and nutrient use efficiency (NUE) is required to achieve the large rise in global food supply and to optimize the performance of our resource-constrained planet. Nearly all farmland soils around the globe suffer from at least one of the vital nutrients required to promote robust plant growth. Soil degradation can be caused by acidity, alkalinity, salinity, human activities, farming practices, and erosion. For an adequate nutrient supply and optimal yields, fertilizer and/or amendment additions are crucial.

Evidence reported an expansion of 200 to 300 percent in the utilization of artificial fertilizer from 1970 to 2010 (FAOSTAT, 2013; Smith et al., 2014). Recent findings confirmed that almost half of the global population depends on nitrogenous inputs to sustain their production (Erisman et al., 2008; Smil, 2002). Climate change, environmental harm, biodiversity loss, urbanization, pesticide residue accumulation, and unbalanced fertilizer use are the main concerns facing today's agriculture. However, due to different factors such as leaching, photolysis, hydrolysis, fixation, and decomposition, most applied nutrients might not be accessible to plants. It is imperative to employ fertilizers in the most efficient manner possible to ensure the optimal growth and development of plants. Consequently, it is vital to reduce nutrient wastage during fertilization and enhance crop yield by utilizing cutting-edge technologies.

6.2 THE NUTRIENT CYCLE

Soil nutrients are lost through several pathways, which are graphically represented in the picture. Higher biomass yields were supported by increased fertilizer (N, P) inputs in the latter part of the 20th century. This is now widely regarded as unsustainable. Alternatives are needed that utilize the soil's natural fertility more efficiently, enhance resource utilization efficiency, and prevent N and P losses (Figure 6.1).

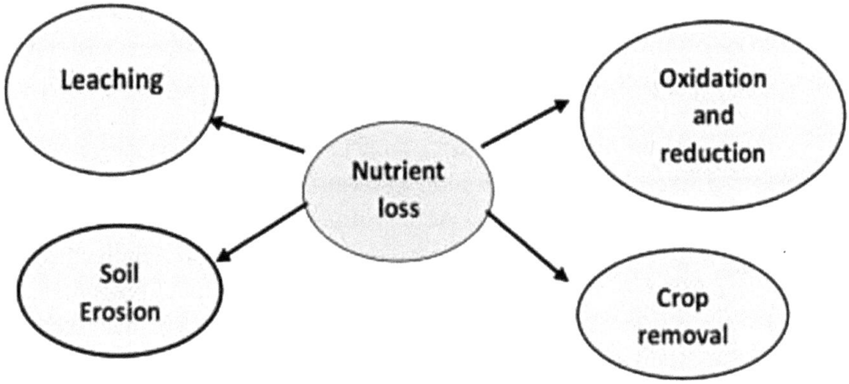

FIGURE 6.1 Various pathway of nutrient loss from soil.

The strategy that focuses on realigning nitrogen and phosphorus inputs, reducing nitrogen and phosphorus wastage in water, minimizing the effects of eutrophication, recycling the N and P found in bio-resources, recovering nitrogen and phosphorus from waste for fertilization purposes, and redefining the use efficiency of phosphorus fertilizer in the agri-food web would lead to more effective nutrient management at the food system level.

In terms of nutrition, plants require the proper ratio of nitrogen (N), phosphorus (P), and potassium (K) in their diet because these substances work together to support plant health, growth, and production (Agyin-Birikorang et al., 2022). Unfortunately, a prior study found that N, P, and K given to the soil were lost at rates of 40–70%, 80–90%, and 50–90%, respectively. This represents a significant loss of important macronutrient resources. Industrialization, deforestation, overgrazing, and other human activities have dramatically decreased the amount of nutrients in the Earth's surface. Nutrient depletion reduces plant improvement and yield, which has a detrimental effect on food security (Ramadan et al., 2020). Chemical fertilizers have been overused in an attempt to retain soil nutrients, but they have severe effects on the environment. Widespread utilization of fertilizers results in the escalation of greenhouse gas emissions along with soil contamination, which leads to the accumulation of pollutants and their transfer through the food chain (Khodakovskaya et al., 2013; Lehemann et al., 2006). Fertilizers are well known for being one of the major contributors to soil and aquatic pollution as nitrates permeate into the soil and groundwater, as well as the discharge of N and P into aquatic bodies (Chien et al., 2009). This has caused soil contamination and negative effects on soil biota in numerous places (Bah et al., 2014).

Apart from the environmental hazards, it is also true that fertilizers are essential to promote faster growth of plants and agriculture. There are two ways to accomplish this goal. The first way is by adding additives that are high in nutrients. Certain fertilizers alter water retention and aeration to boost the soil's efficiency (Roberts, 2009). Hence, understanding how to use less fertilizer is crucial for both sustainable agricultural development and environmental conservation. Achieving increased NUE (Nutrient Use Efficiency) and better agricultural production has become challenging due to the expanding global food demand, dwindling natural resources, and decreasing environmental conditions. Researchers and farmers are now investigating methods for boosting the yield per unit of land. Nevertheless, there are cases where augmenting yield per unit of land exerts stress on the farmland. The soil will be over-harvested of nutrients if the situation is not correctly handled. Increased crop yields and decreased fertilizer input costs are two benefits of improved NUE for plants. One of the best solutions in such circumstances is to incorporate soil amendments. It is essential to find a substitute for a fertilizer that not only minimizes ecosystem pollution but also supplies the nutrients for crop growth. Biochar/nano biochar is one of these materials that also improves the effectiveness of fertilizer. Nowadays, research is being done on nanocomposites made of biochar for their unique features and potential applications. Nano biochar can be the alternative solution for fertilizer.

6.3 WHY NOT BIOCHAR?

Biochar is employed to enhance soil conditions for the betterment of both farming and ecological objectives (Sohi, 2012; Muhammad et al., 2018). Biochar is a carbonaceous material created when plant (or animal) biomass is thermochemically converted. By adding it to the soil, you can increase the amounts of carbon, nitrogen, and phosphorus, reduce phytotoxic effects from contaminated soil, encourage soil microbial ecology, and boost growth and production (Yeboah et al., 2020; Antón-Herrero et al., 2021; Chen et al., 2021; Pan et al., 2021). Biochar is a practical solution for cropping systems due to its nutrient content, larger surface area, porous structure, and potential to serve as a substrate for microflora (Hussain et al., 2017; Muhammad et al., 2018). Furthermore, this material could improve the cation-exchange capacity (CEC) of the soil (Ding et al., 2017) and reduce the global climate crisis. All of these properties make biochar an eco-friendly soil amendment to boost grain output and increase the effectiveness of chemical fertilizers. On the other hand, some researchers assume that biochar's high adsorption capacity is what limits the amount of nutrients that plants can receive, as it closely binds to nutritional elements (Oleszczuk et al., 2014). Also, it may not be viable to employ soil biochar to optimize fertilizer techniques because high rates (tons per hectare) are frequently used, and for that reason, soil biochar may be lost through runoff. As a result, biochar as a nutrient carrier must be energized or stimulated before it can be effectively used in soil layers (Kizito et al., 2019).

6.4 NANO BIO-BIOCHAR

Nano biochar is a substance abundant in carbon that is produced through the pyrolysis process and the carbonization of organic matter. During the carbonization process, micro-sized biochar with dimensions less than a micrometer (μm) and up to a nanometer (nm), referred to as "dissolved" and "nano-biochar," was created (Rajput et al., 2022). In the latest work, nano-biochar with diameters less than 5 nm were derived. In addition to physicochemical qualities, structural modifications also make bulk-Biochar and nano-Biochar (dissolved and nano-biochar) unique from one another. Biochar-based nano compositions are novel materials that combine the benefits of biochar and nanoparticles. Nanoparticles are particles that fall within the size range of 1–100 nm. Their use could provide plants with protection by acting as either insecticides or fertilizers, accelerating plant development, aiding nutrient absorption, and in the case of hydrogels, this will accelerate the effectiveness of water management technology (Jeffery et al., 2016; Li et al., 2017). Moreover, the introduction of nano biochar results in a significant enhancement of its characteristics, including its porosity, chemical bonding, and surface-active sites (Blackwell et al., 2015; Matoso et al., 2019; Farhangi-Abriz & Ghassemi-Golezani, 2021). Nano biochar is characterized by three types of pores: macrospores with diameters exceeding 50 nm, mesopores with diameters ranging from 2 to 50 nm, and microspores. The micrometer-sized pores in nano biochar facilitate the adsorption and desorption of both organic and inorganic ions and molecules, thereby increasing its potential as a soil nutrient holder (Hawthorne et al., 2017; He et al., 2016; Heydari et al., 2016). The characteristics of nano biochar are discussed in Figure 6.2. Owing to its huge specific surface and tiny dimensions, nano-carbon can quickly bond to soil micro-elements and nutritional components, making it an efficient fertilizer. Studies conducted by Mesarič et al. (2013); Guo et al. (2013); and Zhou et al. (2018) have revealed that the use of nano-biochar significantly modifies soil moisture transport and is capable of reducing nutrient loss brought by downpours erosion on loess slopes with changing natural vegetation. Additionally, Wang et al. (2016) have reported that the use of nano-biochar fertilizers boosts crop output by 10% to 20% while simultaneously reducing the usage of fertilizer by 30% to 50%.

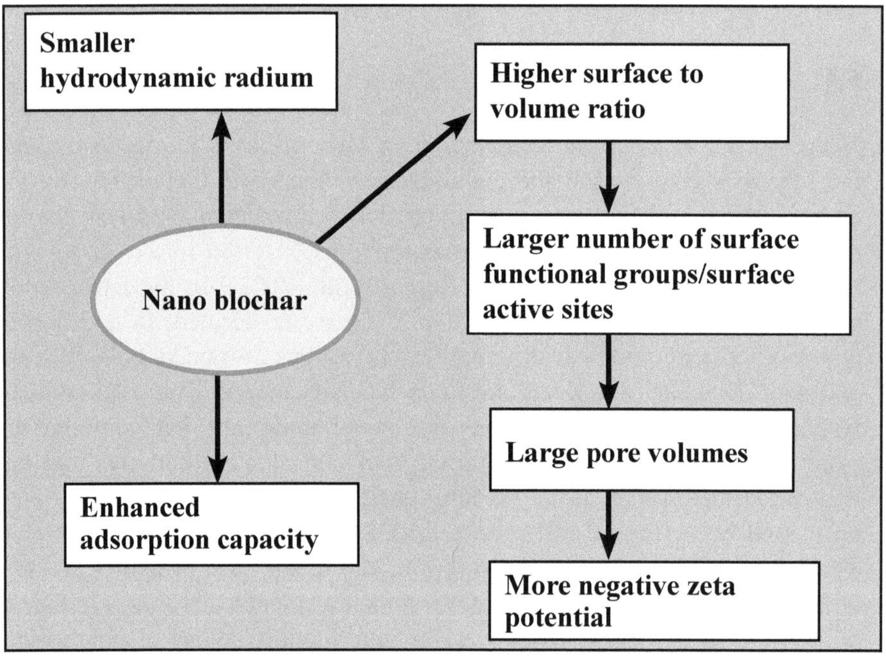

FIGURE 6.2 Characteristics of nano-biochar.

6.5 WHY NANOCOMPOSITES BASED ON BIOCHAR

Innovative biochar-based solutions called nanocomposites combine the benefits of nanoparticles and biochar. The physical and chemical properties of soil, such as permeability, functional groups, and surface-active sites, are significantly enhanced due to the application of nanocomposites (Blackwell et al., 2015; Matoso et al., 2019; Farhangi-Abriz & Ghassemi-Golezani,

2021). Diverse types of hybrid materials based on biochar, such as biochar composites with magnetic properties, composites with nano-metal oxides/hydroxides, and biochar coated with functional nanoparticles, have been put in place to strengthen its suitability (Huang et al., 2019; Li et al., 2020). When biochar is combined with nano-sized magnetic iron oxides such as Fe_3O_4, Fe_2O_3, and $CoFe_2O_4$, the active sites of iron oxide can remove environmental pollutants such as Pb(II) and Cd(II) (Reddy & Lee, 2014). Moreover, these combinations also augment plant performance and tolerance to salt upsurge (Ghassemi-Golezani et al., 2021). The unique structure of biochar-based nanofibers of iron (Fe) and zinc (Zn) has the potential to mitigate salt accumulation, reduce the effect of sodium (Na) absorption, and increase nutrient availability and water status to maximize crop productivity (Ghassemi-Golezani & Rahimzadeh, 2022). The high sodium sorption capacity was the reason for the higher root and shoot K^+/Na^+ ratios of plants treated with nanocomposite materials (Rahimzadeh & Ghassemi-Golezani, 2022).

6.6 DIFFERENT WAYS TO INCREASE THE NUTRIENT AVAILABILITY THROUGH NANO BIO CHAR

6.6.1 IMPACT OF NANO BIOCHAR FOR INCREASING NITROGEN USE EFFICIENCY IN SOIL

The most often used fertilizer is nitrogen, which is a nutrient that limits crop development. By 2022, 111.59 million tons of nitrogen fertilizer are anticipated to be consumed worldwide (FAO, 2019). The utilization of nitrogen-based fertilizers results in significant impurities in the atmosphere, as well as in both surface and groundwater, through the discharge of ammonia (NH_3), leaching of nitrate (NO_3), and the release of nitrous oxide (N_2O) (Wen et al., 2017). Crops and plants only absorb 20–30% of the nitrogen present in fertilizers (Naz et al., 2016). Nano-biochar slows down the rate of nitrogen movement into the deeper layers of soil (Albiter et al., 2010). Nitrate builds up in the soil layers because of excessive nitrogen fertilization, which also raises the possibility of leaching in the form of nitrate (Zhang et al., 2010). According to Dempster et al. (2016), using biochar can speed up nitrate adsorption. Hence, in fertilized soils, banding nano-biochar at a depth of 5 to 10 cm can prevent the leaching of nitrate and improve nitrogen use efficiency.

6.6.2 IMPACT OF NANO-BIOCHAR IN INCREASING SOIL PHOSPHORUS AVAILABILITY

When it comes to agricultural production, phosphorus is crucial for crop growth. By 2022, there will be a demand of 49.10 million tons for phosphorus fertilizers worldwide (FAO, 2019). The finite supply of phosphorus rock is running out. Based on supply and demand dynamics for phosphorus, the world's high-quality phosphate stocks will be used up within 50–400 years (van Dijk et al., 2016). The inadequate accessibility of soluble phosphorus for plant absorption creates difficulties with phosphorus fertilizers. As phosphorus is poorly soluble, plants can only absorb it when it is present in the inorganic forms of HPO_4^{2-}, H_2PO_4, and H_3PO_4. Due to interactions with the soil, including adsorption, precipitation, and conversion into the organic form, phosphorus is highly immobile. The primary cause of eutrophication is the loss of phosphorus into surface water, which leads to soil pollution. When nano-biochar was mixed with soil, it enhanced the soil's ability to retain accessible phosphorus in the upper layers. Phosphorus is highly stable and has low mobility, which makes it easily absorbed by soil particles (Sharma et al., 2015; Petr et al., 2011). The application of more biochar resulted in an increase in the amount of soil-available phosphorus (Arif et al., 2017). The addition of nano-biochar can significantly enhance the physical and chemical characteristics of soil by increasing the proportion of larger particles and decreasing soil porosity. The absorption of phosphorus by soil particles was substantial in the rhizosphere where the application of nano-biochar was blended, which leads to its stabilization. Because of this, spring maize had greater availability to phosphorus in the root zone (Yang et al., 2017).

6.6.3 IMPACT OF NANO-BIOCHAR IN INCREASING SOIL POTASSIUM AVAILABILITY

The accumulation of nano-biochars on the upper layer of soil particles promotes an upsurge in the count of microscopic soil pores. Nevertheless, water-soluble potassium ions tend to sink into the soil because of the poor soil adsorption of potassium ions in layers of soil. Upon the application of nano-biochar on the soil, it obstructs potassium ions from migrating in a straight line. As a result, the root layer has elevated potassium ion concentrations and the attainable potassium ions remain in the nano-biochar layer for an extended span. This augments the utilization of potassium by crops. In accordance with a particular study, during the tasselling stage, the standard obtainable potassium level was enhanced at rates of 0.7% and 1% escalated by 11.1% and 22.6%, respectively when blended with nano-biochar soil as compared to control (Liu et al., 2012). Also, Liu et al. (2012) discovered that the addition of nano-biochar to compost amplified the quantities of plant-accessible Ca, K, P, and Na by a factor of 2.2, 2.5, 1.2, and 2.8, respectively. Nano-biochar can boost the inherent fertility of the soil by enriching the soil's availability to micronutrients such as potassium (Dong et al., 2016).

6.6.4 ENHANCING SOIL ORGANIC CARBON CONTENT IN SOIL THROUGH NANO-BIOCHAR

Organic matter in soil, commonly referred to as soil organic carbon (SOC), plays a pivotal role in maintaining soil fertility and overall soil well-being. Recent research has shown that the addition of nano biochar significantly accelerates the build-up of organic carbon content in soil (Kianfar, 2020; Syah et al., 2021; Majdi et al., 2021). For example, incorporating 8% of nano biochar into soil results in a substantial 41% increase in organic carbon levels, which remain stable for up to 210 days without any reduction in the amount of nano biochar added. Furthermore, nano biochar has been found to intensify the long-term stability of soil organic carbon and reduce its mineralization, as per the research.

6.6.5 IMPROVING BULK DENSITY AND WATER HOLDING CAPACITY IN SOIL THROUGH NANO-BIOCHAR

Soil with a lower bulk density promotes root growth, nutrient availability, aeration, microbial population, and water-holding capacity. Incorporating nano biochar into the soil enhances its porosity while reducing its bulk density, leading to an improved capacity to retain water and an increase in nutrient accessibility. According to research, the combined use of nano-Fe_3O_4 and biochar can boost the overall pore volume and micropore volume by approximately 60 and 6 times, respectively (Zhang et al., 2020).

6.6.6 IMPACT OF USING NANO-BIO CHAR AS FERTILIZER

Biochar-derived nanocarbons have piqued the interest of researchers due to their ability to store water and nutrients (Manya, 2012; Xiu et al., 2017) and can be used as nano-fertilizers (Rose et al., 2011; Tiwari et al., 2014). Macronutrients including nitrogen (N), phosphorus (P), magnesium (Mg), potassium (K), calcium (Ca), and sulfur (S), as well as micronutrients like copper (Cu), manganese (Mn), iron (Fe), zinc (Zn), and ash, are also present in elevated amounts in carbon-rich biochar made from natural biomass (Hossain, 2011), making it a high-performance organic fertilizer. According to Bop et al. (2016), biochar derived from animal feed sources such as broiler litter and sewage sludge exhibited elevated levels of both phosphorus and nitrogen. Lateef et al. (2019) recently explained whether a biochar nano-composite made from maize cobs helped to conserve the environment by transforming trash into a profitable product and solved the leaching concerns associated with traditional fertilizer. Mixing biochar with other fertilizers, such as urea, resulted in a significant increase in yield while reducing N input (Joseph et al., 2013; Manya, 2012). Additionally, the utilization of fertilizers can be enhanced by incorporating minerals and organic substances into biochar composites. Furthermore, Zhang et al. (2020) demonstrated that the utilization of carbon-based nanoparticles significantly boosted wheat crop development and productivity. There are various nanocarbons and nanobio-char used for plant growth promotion which are detailed in Table 6.1.

TABLE 6.1 Examples of Nanocarbon and Nano-Biochar Which Is Used as Fertilizer

Nanocarbons/ nano-biochar as feertilizers	Plant	Outcome	References
Biochar derived nano carbon fertilizer	Rice, Gram,Tomato	Production of crops	Khodakovskaya et al. (2013); Lin et al. (2009)
Water-soluble carbon nanoparticles CNPs (ws-CNPs)	Wheat	Improved root length, as well as seedling growth	Sexena et al. (2014)
Liquid nanocarbon	Chinese cabbage	Growth and output of crops.	Zhao et al. (2017)
Nano carbon water retaining fertilizer	Tuber mustard	Better improved quality, yield, and fertilizer input efficiency	Wang et al. (2018)

Sources: Modified from Chausali et al., (2021).

6.6.7 ENHANCEMENT OF MICROBIAL COMMUNITY IN SOIL THROUGH NANO-BIOCHAR

Microbes are the best players for nutrient cycling in the soil. The existence of diverse organic substances has been the primary cause for the ideal habitat of microorganisms, which is advantageous for the soil ecosystem in numerous ways. Microbial consortium plays an important role in the well-functioning of crop soil ecosystems by preserving the inherent fertility of soil, nutrient cycling, improving soil health condition, inhibiting crop disease, and boosting crop yields (Thies et al., 2015). Biochar has been demonstrated to alter the physical and chemical soil parameters such as surface area, porosity, bulk density, tensile strength, pH, CEC, among others (Mangrich et al., 2015; Chan et al., 2007), which subsequently modifies soil biota and has a substantial effect on soil microbial populations (Lehman et al., 2011).

As per the study conducted by Liu et al. in 2020, the utilization of nano-biochar treatment resulted in a noteworthy rise in microbial diversity by increasing the population of Actinobacteria and Bacteroidetes in Cd-polluted soil. At the same time, it led to a reduction in the number of Proteobacteria, which is usually dominant in contaminated areas. These findings were further supported by the research conducted by Liao et al. in 2019. Additionally, Wu et al. (2020) introduced a novel calcium-based magnetic biochar (Ca-MBC) that remarkably amplified the count of the bacterial population and transformed the ratio of microbial taxa.

6.6.8 IMPACT OF NANO-BIOCHAR FOR ENHANCING SOIL FERTILITY

Biochar that has been supplemented with nutrients has the potential to create a distinctive soil-plant-environment system to practice sustainable agriculture (Rizhiya et al., 2020). Biochars fortified with macronutrients, particularly nitrogen and phosphorus, hold promise in enhancing soil productivity, lessening the requirement for fertilizers, and repurposing nutrients obtained from agricultural leftovers or organic leftovers (Chia et al., 2014; Moradi et al., 2019; Karim et al., 2021). The enhanced biochar augmented phosphorus, iron, and zinc levels while dropping soil pH by approximately 1.4 percent. An altered biochar with elevated nitrogen content raises the amount of nitrogen in the soil overall (Chan et al., 2008; Sarkhot et al., 2012; Moradi et al., 2019) and reduces CO_2 and N_2O emissions from the soil (Rizhiya et al., 2020). The use of enriched biochar with appetite and phosphoric acid measured a high amount of bioavailable phosphorus in the soil, and it accelerates the phosphate solubilizing bacteria in the soil (Chia et al., 2014). Because of the good effects on soil quality associated with greater organic matter content, primary nutrients, micro-nutrient, and heavy metal immobilization, fortified biochar or modified biochar may substitute chemical fertilizers and support organic farming (Kizito et al., 2019) and provide its beneficial effects on soil fertility. The utilization of altered biochars in soils lacking in nutrients can alter the process of nutrient recycling, leading to an increase in crop productivity. The oxygenated functional groups present in modified biochar facilitate the development of negative surface charge and greater nutrient adsorption process (Tan et al., 2020). Chickpea growth and yield were significantly enhanced by adding phosphorus-enriched biochar to the soil, which improved plant nodulation process and physiological function (Wali et al., 2020). Positive effects on biochemical properties and soil fertility were seen when urea ammonium nitrate added nitrogen to biochar. The overall growth and productivity of *Zea mays* L. improved significantly owing to elevated nitrogen accessibility and uptake, decreased C:N ratio, and heightened efficiency in nitrogen usage by the plant (Dil et al., 2014). Research carried out to showcase the effects of nitrogen-enriched biochar (fortified with ammonium sulfate) on the growth of *Oryza sativa* L. or rice plants demonstrated that using nano-biochar as a favorable medium for nitrogen in crop nourishment (Utomo et al., 2017). In order to maximize the effectiveness of nitrogen fertilization, enriched biochar has higher cation exchange capacity, which lowers the denitrification and leaching loss of this nutrient (Utomo et al., 2017). Yeboah et al. (2017) claim that the addition of nitrogen-enriched biochar boosted spring wheat grain yield through increasing water content in soil and photosynthetic capabilities in plants. Wen et al. (2017) found that enhanced biochar containing NH_4^+ acts as a slow-release nitrogen fertilizer which enhances cotton plant development through boosting the retention of soil water and nitrogen-use efficiency. In safflower plants, sodium build-up and salt toxicity were minimized when MgO and MnO nanocomposites based on biochar were added to the soil (Ghassemi-Golezani et al., 2021). Such studies were related to elevated concentrations of chlorophyll activity and electron transport pathways, which improved plant development by inhibiting oxidative stress. In a salinity environment, biochar-based nanocomposites stimulated the development of crops by lowering the exchangeable sodium percentage (ESP) and sodium adsorption ratio (SAR) in problematic soil (Farhangi-Abriz & Ghassemi-Golezani, 2021; Rahimzadeh & Ghassemi-Golezani, 2022). Increasing the rate of nutrient absorption can supply nutrients to plant cells, improve the water status of plant tissues, foster better plant development, and enhance the ability of plants to withstand stress (Farhangi-Abriz & Ghassemi-Golezani, 2021). Joseph et al. (2015) demonstrated the utilization of biochar generated from Acacia saligna wood and reinforced with magnetic iron nanoparticles (Fe_3O_4, γ-Fe_2O_3), which led to an augmentation in overall plant development.

There was an increase in the absorption of P and N by the crops, partially linked to an improvement in mycorrhizal colonization and enhancement of soil biology in wheat crops.

6.6.9 ENHANCEMENT OF SOIL CHEMICAL PROPERTIES THROUGH NANO-BIOCHAR

The properties of recognized and unknown natural substances in biochar modify as time goes on. This occurs because of actions such as exposure to air, contact with natural substances and soil minerals, and decomposition by microorganisms. The inclusion of biochar results in modifications in pH, electrical conductivity (EC), cation exchange capacity (CEC), and soil nutrient content. The pH levels of the biochar can vary significantly, ranging from 4 to 12, depending on the source material used. The functional groups in nano biochar have been implicated in both the pH increase and drop. According to researchers, the fundamental reason for the pH decline, according to studies, is the oxidation of carbon to carboxylic acid groups. While alkaline substances' dissolution is thought to be what causes the pH to rise (Rhoades et al., 2017; Wang et al., 2017). Consequently, with the passage of time, the presence of biochar in the soil can modify the soil's pH level. At first, biochar exhibits a lower capacity for exchanging cations compared to organic matter in the soil, and subsequently, it would carry either a positive or negative overall electrical charge (Mohammad et al., 2021; Saied et al., 2021; Kundu et al., 2020). The functional groups of organic acids like humic acid are primarily related to the CEC in biochar, which are eliminated at higher formation temperatures. The use of high CEC biochar has the potential to improve soil quality and minimize nitrogen loss caused by sedimentation, adsorption to soil minerals, and leaching. Research indicates that biochar can reduce the leaching rate of nitrogen-based fertilizers (such as ammonium compounds) by approximately 60%, increasing the fertilizer's effectiveness and supplying more nutrients to the soil. Moreover, soil organic matter can be absorbed by biochar, conserving soil organic carbon while also boosting its CEC. This may be the reason why soil organic matter levels increase over time (Honarvar et al., 2021; Aramesh Boroujeni & Asadi Aghbalaghi, 2021) and soil organic carbon does not degrade. Because of its ability to exchange cations with other elements, biochar can effectively reduce the saltiness (EC) and osmotic potential of soil solutions by absorbing substances including sodium, calcium, and magnesium (Wrobel et al., 2016; Xu et al., 2018;

Yuan et al., 2017). The plant also experiences reduced salt stress. Moreover, biochar can lower the level of soluble substances like phenol in the soil solution. Aluminum and manganese are two elements whose concentration in the soil solution rises in acidic soils (Zhang et al., 2016). Flora and microbes have the ability to make use of biochar since specific components, including Silicon, Iron, Sulphur, Phosphorus, Potassium, Magnesium, and Calcium, are liberated from it and infiltrate the soil solution once it is blended with acidic solutions (Hua et al., 2014; Peterson et al., 2013; Ulyett et al., 2014) and increase the nutrient availability.

6.6.10 IMPACT OF NANO-BIOCHAR ON PLANT NUTRIENTS

When being added to the soil, nano-biochar significantly adsorbs soil nutrients, preventing nitrogen from moving vertically. As a result, nitrogen is moved from the roots to other plant components, increasing the nitrogen levels in the plants. This encourages crops to absorb nitrogen from the soil. The study also noticed that at various growth stages, soil amended with nano-biochar had a greater level of phosphorus content compared to the control group in the spring maize group. Yang et al. (2020) show that the plots growing spring maize altered with nano-biochar had significant levels of nitrogen.

6.7 CONCLUSION

Biochar exhibits several distinct features, including a wide surface area, porous microstructure, graphitic composition, and an abundance of surface groups. These properties make it a highly desirable material for a variety of applications. One promising use for biochar is in soil improvement. Nano biochar, in particular, has been shown to enhance water and fertilizer efficiency and can serve as a substrate for macro- and micro-nutrients, ensuring that all nutrients are absorbed by the substrate. Additionally, nano biochar can promote plant growth in saline soils and can effectively remove organic and inorganic contaminants from soil. By using materials such as nano biochar, we can protect fertile farmland from destruction and erosion while increasing agricultural production per unit area. This sustainable alternative to harmful chemical fertilizers can also help reduce the greenhouse effect, making it a crucial component of environmentally conscious agriculture.

KEYWORDS

- **biochar**
- **cation exchange capacity**
- **electrical conductivity**
- **nano-biochar**
- **pH**
- **soil nutrient content**

REFERENCES

Agyin-Birikorang, S., Tindjina, I., Adu-Gyamfi, R., Dauda, H. W., Fugice Jr, J., & Sanabria, J., (2022). Managing essential plant nutrients to improve maize productivity in the savanna agroecological zones of northern Ghana: The role of secondary and micronutrients. *Journal of Plant Nutrition, 46*(1), 38–57.

Albiter, M. A., Crooks, R. M., Zaera, F., et al., (2010). Adsorption of carbon monoxide on dendrimer-encapsulated platinum nanoparticles: Liquid versus gas phase. *The Journal of Physical Chemistry Letters, 1*(1), 38–40.

Antón-Herrero, R., Vega-Jara, L., García-Delgado, C., Mayans, B., Camacho-Arévalo, R., Moreno-Jiménez, E., Plaza, C., & Eymar, E., (2021). Synergistic effects of biochar and biostimulants on nutrient and toxic element uptake by pepper in contaminated soils. *J. Sci. Food Agric., 102*, 167–174.

Aramesh, B. Z., & Asadi, A. Z., (2021). Electrochemical determination venlafaxine at NiO/GR nanocomposite modified carbon paste electrode. *Iranian Journal of Chemistry and Chemical Engineering (IJCCE), 40*(4), 1042–1053. doi: 10.30492/ijcce.2020.39188.

Arif, M., Ilyas, M., Riaz, M., et al., (2017). Biochar improves phosphorus use efficiency of organic-inorganic fertilizers, maize-wheat productivity and soil quality in a low fertility alkaline soil. *Field Crops Research, 214*, 25–37.

Bah, A., Husni, M. H. A., Teh, C. B. S., et al., (2014). Reducing runoff loss of applied nutrients in oil palm cultivation using controlled-release fertilizers, *Adv. Agric.*, 285387.

Baulcombe, D., Crute, I., Davies, B., Dunwell, J., Gale, M., Jones, J., Pretty, J., et al., (2009). *Reaping the Benefits: Science and the Sustainable Intensification of Global Agriculture.* The Royal Society, London, United Kingdom.

Blackwell, P., Joseph, S., Munroe, P., Anawar, H. M., Storer, P., Gilkes, R. J., & Solaiman, Z. M., (2015). Influences of biochar and biochar-mineral complex on mycorrhizal colonization and nutrition of wheat and sorghum. *Pedosphere, 25*, 686–695. doi: 10.1016/S1002-0160(15)30049-7.

Bopp, C., Christl, I., Schulin, R., & Evangelou, M. W., (2016). Biochar as possible long–term soil amendment for phytostabilization of TE–contaminated soils. *Environ. Sci. Pollut. Res., 23*, 17449–17458.

Chan, K. Y., Van, Z. L., Meszaros, I., Downie, A., & Joseph, S., (2008). Using poultry litter biochars as soil amendments. *Australian Journal of Soil Research, 46*, 437–444. doi: 10.1071/SR08036.

Chan, K. Y., Zwieten, L. V., Meszaros, I., Downie, A., & Joseph, S., (2007). Agronomic values of green waste biochar as a soil amendment. *Aust. J. Soil Res., 45*, 629–634.

Chausalia, N., Saxena, J., & Ram, P., (2021). Nanobiochar and biochar-based nanocomposites: Advances and applications. *Journal of Agriculture and Food Research, 5*, 100191.

Chen, X., Yang, S., Ding, J., Jiang, Z., & Sun, X., (2021). Effects of biochar addition on rice growth and yield under water-saving irrigation. *Water, 13*, 209.

Chia, C. H., Singh, B. P., Joseph, S., Graber, E. R., & Munroe, P., (2014). Characterization of an enriched biochar. *Journal of Analytical and Applied Pyrolysis, 108*, 26– 34.

Chien, S. H., Prochnow, L. I., & Cantarella, H., (2009). Chapter 8 recent developments of fertilizer production and use to improve nutrient efficiency and minimize environmental impacts. *Adv. Agron., 102*, 267–322.

Dempster, D. N., Jones, D. L., & Murphy, D. V., (2012). Clay and biochar amendments decreased inorganic but not dissolved organic nitrogen leaching in soil. *Soil Research, 50*(3), 216–221.

Dil, M., Oelbermann, M., & Xue, W., (2014). An evaluation of biochar pre-conditioned with urea ammonium nitrate on maize (*Zea mays* L.) production and soil biochemical characteristics. *Canadian Journal of Soil Science, 94*, 551–562. doi: 10.4141/cjss-2014-010.

Ding, Y., Liu, Y., Liu, S., Huang, X., Li, Z., Tan, X., Zeng, G., & Zhou, L., (2017). Potential benefits from biochar application for agricultural use: A review. *Pedosphere, 27*, 645–661. doi: 10.1016/S1002-0160(17)60375-8.

Dong, X., Guan, T., Li, G., et al., (2016). Long-term effects of biochar amount on the content and composition of organic matter in soil aggregates under field conditions. *Journal of Soils and Sediments, 16*(5), 1481–1497.

Erisman, J. W., Sutton, M. A., Galloway, J., Klimont, Z., & Winiwarter, W., (2008). How a century of ammonia synthesis changed the world. *Nat. Geosci., 1*, 636–639. https://doi.org/10.1038/ngeo325.

FAO (2009). FAOSTAT. *FAO Statistics Division*. Online at: http://faostat.fao.org/ (accessed on 25 April 2024).

FAO (2013). *FAOSTAT Database*. Food and Agriculture Organization of the United Nations, Rome, Italy.

FAO, (2019). *World Fertilizer Trends and Outlook to 2022: Summary Report.*

Farhangi-Abriz, S., & Ghassemi-Golezani, K., (2021). Changes in soil properties and salt tolerance of safflower in response to biochar-based metal oxide nanocomposites of magnesium and manganese. *Ecotoxicology and Environmental Safety, 211*, 111904. doi: 10.1016/j.ecoenv.2021.111904.

Ghassemi-Golezani, K., & Rahimzadeh, S., (2022). Biochar-based nutritional nanocomposites: A superior treatment for alleviating salt toxicity and improving physiological performance of dill (*Anethum graveolens*). *Environmental Geochemistry and Health*, PMID: 36153765. doi: 10.1007/ s10653-022-01397-4.

Ghassemi-Golezani, K., Farhangi-Abriz, S., & Abdoli, S., (2021). How can biochar-based metal oxide nanocomposites counter salt toxicity in plants. *Environmental Geochemistry and Health, 43*, 2007–2023. doi: 10.1007/ s10653-020-00780-3.

Glenn, J. C., Gordon, T. J., & Florescu, E., (2008). *The Millenium Project: State of the Future.* World Federation of UN Associations, Washington, DC.

Guo, Y., Li, Y., Zhu, T., & Ye, M., (2013). Effects of concentration and adsorption product on the adsorption of SO_2 and NO on activated carbon. *Energy Fuels, 27*(1), 360–366.

Hawthorne, I., Johnson, M. S., Jassal, R. S., Black, T. A., Grant, N. J., & Smukler, S. M., (2017). Application of biochar and nitrogen influences fluxes of CO_2, CH_4 and N_2O in a forest soil. *J. Environ. Manag., 192*, 203–214.

He, Y. H., Zhou, X. H., Jiang, L. L., Li, M., Du, Z. G., Zhou, G. Y., Shao, J. J., et al., (2016). Effects of biochar application on soil greenhouse gas fluxes: A meta-analysis. *GCB Bioenergy, 9*, 743–755. 25.

Heydari, M., Prévosto, B., Naji, H. R., Mehrabi, A. A., & Pothier, D., (2017). Influence of soil properties and burial depth on Persian oak (*Quercus brantii*, Lindl) establishment in different microhabitats resulting from traditional forest practices. *Eur. J. Forest Res., 136*, 1–19.

Honarvar, Z., Farhoodi, M., Khani, M., Mohammadi, A., Shokri, B., Jannatiha, N., & Shojaee-Aliabadi, S., (2021). Antimicrobial and physicochemical properties of plasma treated bio-coating polypropylene films containing *Satureja hortensis* essential oil. *Iranian Journal of Chemistry and Chemical Engineering (IJCCE), 40*(4), 1216–1228. doi: 10.30492/ijcce.2020.43324 122.

Hossain, M. K., Strezov, V., Chan, K. Y., Ziolkowski, A., & Nelson, P. F., (2011). Influence of pyrolysis temperature on production and nutrient properties of wastewater sludge biochar. *J. Environ. Manag., 92*, 223–228.

Hua, L., Lu, Z., Ma, H., & Jin, S., (2014). Effect of biochar on carbon dioxide release, organic carbon accumulation, aggregation of soil. *Environmental Progress & Sustainable Energy, 33*(3), 941–946.

Huang, Q., Song, S., Chen, Z., Hu, B., Chen, J., & Wang, X., (2019). Biochar-based materials and their applications in removal of organic contaminants from wastewater: State of-the-art review. *Biochar, 1*, 45–73. doi: 10.1007/s42773- 019-00006-5.

Hussain, M., Farooq, M., Nawaz, A., Al-Sadi, A. M., Solaiman, Z. M., Alghamdi, S. S., Ammara, U., Ok, Y. S., & Siddique, K. H. M., (2017). Biochar for crop production: Potential benefits and risks. *Journal of Soils and Sediments, 17*, 685–716. doi: 10.1007/s11368-016-1360-2.

Jeffery, S., Verheijen, F. G., Kammann, C., & Abalos, D., (2016). Biochar effects on methane emissions from soils: A meta-analysis. *Soil Biol. Biochem., 101*, 251–258.

Joseph, S., Anawar, H., Storer, P., Blackwell, P., Chia, C., Munroe, P., Donne, S., et al., (2015). Effects of enriched biochars containing magnetic iron nanoparticles on mycorrhizal colonization, plant growth, nutrient uptake and soil quality improvement. *Pedosphere, 25*, 749–760. doi: 10.1016/S1002- 0160(15)30056-4.

Joseph, S., Graber, E. R., Chia, C., et al., (2013). Shifting paradigms: Development of high–efficiency biochar fertilizers based on nano-structures and soluble components. *Carbon Manag., 4*, 323–343.

Karim, A. A., Kumar, M., Singh, E., Kumar, A., Kumar, S., Ray, A., & Kumar, D. N., (2021). Enrichment of primary macronutrients in biochar for sustainable agriculture: A review. *Critical Reviews in Environmental Science and Technology, 52*(9), 1449–1490. doi: 10.1080/10643389.2020.1859271.

Khodakovskaya, M. V., De Silva, K., Biris, A. S., Dervishi, E., & Villagarcia, H., (2012). Carbon nanotubes induce growth enhancement of tobacco cells. *ACS Nano, 6*(3), 2128–2135.

Khodakovskaya, M. V., Kim, B. S., Kim, J. N., et al., (2013). Carbon nanotubes as plant growth regulators: Effects on tomato growth, reproductive system, and soil microbial community. *Small, 9*, 115–123.

Kianfar, E., (2020). Magnetic nanoparticles in targeted drug delivery: A review. *Journal of Superconductivity and Novel Magnetism.* https://doi.org/10.1007/s10948-021-05932-9. 105.

Kizito, S., Luo, H., Lu, J., Bah, H., Dong, R., & Wu, S., (2019). Role of nutrient-enriched biochar as a soil amendment during maize growth: Exploring practical alternatives to recycle agricultural residuals and to reduce chemical fertilizer demand. *Sustainability, 11*, 3211. doi: 10.3390/ su11113211.

Kundu, R., Biswas, C., Ahmed, J., Naime, J., & Ara, M., (2020). A study on the adsorption of cadmium (II) from aqueous solution onto activated carbon originated from *Bombax ceiba* fruit shell. *Journal of Chemical Health Risks, 10*(4), 243–252. doi: 10.22034/ jchr.2020.1903764.1154.

Lateef, A., Nazir, R., Jamil, A., Nalam, S., Shah, R., Khan, M. N., & Saleem, M., (2019). Synthesis and characterization of environmentally friendly corncob biochar-based nanocomposite–A potential slow-release nano-fertilizer for sustainable agriculture. *Environ. Nanotechnol. Monitor Manag., 11*, 100212.

Lehmann, J., (2007). A handful of carbon. *Nature, 447*, 143, 144.

Lehmann, J., Rillig, M. C., Thies, J., Masiello, C. A., Hockaday, W. C., & Crowley, D., (2011). Biochar effects on soil biota – A review. *Soil Biology Biochemistry, 43*(9), 1812–1836.

Li, J., Cai, X., Liu, Y., Gu, Y., Wang, H., Liu, S., Liu, S., et al., (2020). Design and synthesis of a biochar-supported nano manganese dioxide composite for antibiotics removal from aqueous solution. *Frontiers in Environmental Science, 8*, 62. doi: 10.3389/fenvs.2020.00062.

Li, Z. G., Gu, C. M., Zhang, R. H., Ibrahim, M., Zhang, G. S., Wang, L., Zhang, R. Q., et al., (2017). The benefic effect induced by biochar on soil erosion and nutrient loss of slopping land under natural rainfall conditions in central China. *Agric. Water Manag., 185*, 145–150.

Liao, H., Li, Y., & Yao, H., (2019). Biochar amendment stimulates utilization of plant-derived carbon by soil bacteria in an intercropping system. *Front. Microbiol., 10*, 1361.

Lin, S., Reppert, J., Hu, Q., et al., (2009). Uptake, translocation, and transmission of carbon nanomaterials in rice plants. *Small, 5*, 1128–1132.

Liu, W., Li, Y., Feng, Y., et al., (2020). The effectiveness of nanobiochar for reducing phytotoxicity and improving soil remediation in cadmium-contaminated soil. *Sci. Rep., 10*, 858.

Majdi, H. S., Latipov, Z. A., Borisov, V., et al., (2021). Nano and battery anode: A review. *Nanoscale Res. Lett., 16*, 177. https://doi.org/10.1186/s11671-021-03631-x.

Mangrich, A. S., Cardos, E. M. C., Doumer, M. E., Romao, L., Vidal, L., Rigol, A., & Novotny, E., (2015). Improving the water holding capacity of soils of Northeast Brazil by biochar augmentation,. *ACS Symp. Ser., 1206*, 339–354.

Manya, J. J., (2012). Pyrolysis for biochar purposes: A review to establish current knowledge gaps and research needs. *Environ. Sci. Technol., 46*, 7939–7954.

Matoso, S. C. G., Wadt, P. G. S., Souza, V. S. D., & Pérez, X. L. O., (2019). Synthesis of enriched biochar as a vehicle for phosphorus in tropical soils. *Acta Amazonica, 49*, 268–276.

Mesarič, T., Baweja, L., Drašler, B., et al., (2013). Effects of surface curvature and surface characteristics of carbon-based nanomaterials on the adsorption and activity of acetylcholinesterase. *Carbon, 62*, 222–232.

Mohamed, M. A., Hashim, A. F., Alghuthaymi, M. A., & Abd-Elsalam, K. A., (2018). Nano-carbon: Plant growth promotion and protection, in: Abd-Elsalam, K. A., & Prasad, R., (eds.), *Nanobiotechnology Applications in Plant Protection* (pp. 155–188). Springer.

Mohammad, M., Ahmed, S., & Alameri, R., (2021). Green medicine: A novel preparation method for green synthesizing of iron nanoparticles derived from beta vulgaris extract. *Archives of Razi Institute, 76*(5), 1327–1332. doi: 10.22092/ari.2021.355933.1740 119.

Moradi, S., Rasouli-Sadaghiani, M. H., Sepehr, E., Khodaverdiloo, H., & Barin, M., (2019). Soil nutrients status affected by simple and enriched biochar application under salinity conditions. *Environmental Monitoring and Assessment, 191*(4), 257. doi: 10.1007/s10661-019-7393-4.

Naz, M. Y., & Sulaiman, S. A., (2016). Slow-release coating remedy for nitrogen loss from conventional urea: A review. *J. Contr. Release, 225*, 109–120.

OECD/FAO, (2018). *OECD-FAO Agricultural Outlook 2018–2027*. OECD Publishing Paris/Food and Agriculture Organization of United Nations, Rome. ISBN – 9789251305010 https://doi.org/10.1787/agr_outlook-2018-en.

Oleszczuk, P., Jośko, I., Kuśmierz, M., Futa, B., Wielgosz, E., Ligeza, S., & Pranagal, J., (2014). Microbiological, biochemical and eco-toxicological evaluation of soils in the area of biochar production in relation to polycyclic aromatic hydrocarbon content. *Geoderma, 213*, 502–511. doi: 10.1016/j.geoderma.2013.08.027.

Pan, S. Y., Dong, C. D., Su, J. F., Wang, P. Y., Chen, C. W., Chang, J. S., Kim, H., et al., (2021). The role of biochar in regulating the carbon, phosphorus, and nitrogen cycles exemplified by soil systems. *Sustainability, 13*, 5612.

Peterson, S. C., Jackson, M. A., & Appell, M., (2013). Biochar. *Sustainable Versatile, 1143*, 193–205.

Petr, K., Bernd, S., Astrid, R., et al., (2011). Citramalic acid and salicylic acid in sugar beet root exudates solubilize soil phosphorus. *BMC Plant Biology, 11*(1), 121.

Rahimzadeh, S., & Ghassemi-Golezani, K., (2022). Biochar based nutritional nanocomposites altered nutrient uptake and vacuolar H+ -pump activities of dill under salinity. *Journal of Soil Science and Plant Nutrition, 22,* 3568 –3581. doi: 10.1007/s42729-022-00910-z.

Rajput, V. D., Minkina, T., Ahmed, B., Singh, V. K., Mandzhieva, S., et al., (2022). Nano-biochar: A novel solution for sustainable agriculture and environmental remediation. *Environmental Research, 210*, 112891.

Ramadan, M. M., Amal, A., & Abd-Elsalam, K. A., (2020). Micro/nano biochar for sustainable plant health: Present status and future prospects. In: Abd-Elsalam, A. K., (ed.), *Carbon Nanomaterials for Agri-Food and Environmental Applications* (pp. 323–357). Elsevier BV.

Reddy, D. H. K., & Lee, S. M., (2014). Magnetic biochar composite: Facile synthesis, characterization, and application for heavy metal removal. *Colloids and Surfaces A: Physicochemical and Engineering Aspects, 454*, 96–103. doi: 10.1016/j.colsurfa.2014.03.105.

Rhoades, C. C., Minatre, K. L., Pierson, D. N., Fegel, T. S., Cotrufo, M. F., & Kelly, E. F., (2017). Examining the potential of forest residue-based amendments for post-wildfire rehabilitation in Colorado, USA. *Scientifica*. https://doi.org/10.1155/2017/4758316.

Rizhiya, E. Y., Horák, J., Šimanský, V., & Buchkina, N. P., (2020). Nitrogen enriched biochar-compost mixture as a soil amendment to the haplic luvisol: Effect on greenhouse gas emission. *Biologia., 75*, 873– 884. doi: 10.2478/ s11756-019-00335-7.

Roberts, T. L., (2009). The role of fertilizer in growing the world's food. *Better Crops Plant Food, 93*, 12–15.

Rose, P., Bhat, M., Vidhani, K., Ajmeri, N., Gole, A., & Ghaisas, S., (2011). Intelligent informatics platform for nano–agriculture. In: *2011 11th IEEE International Conference on Nanotechnology* (pp. 916–919). IEEE.

Saied, S., Mohammed, S., Khaleel, B., & Saleh, M., (2021). Comparative studies between conventional techniques and green chemistry to synthesis of novel piperidinium salts ionic liquids (PBSILs). *Journal of Chemical Health Risks, 11*(4), 451–456. doi: 10.22034/jchr.2021.686640 120.

Sarkhot, D. V., Berhe, A. A., & Ghezzehei, T. A., (2012). Impact of biochar enriched with dairy manure effluent on carbon and nitrogen dynamics. *Journal of Environment Quality, 41*, 1107–1114. doi: 10.2134/jeq2011.0123.

Saxena, M., Maity, S., & Sarkar, S., (2014). Carbon nanoparticles in 'biochar boost wheat (*Triticum aestivum)* plant growth. *Res. Adv., 4*, 39948–39954.

Sharma, R., Bell, R. W., & Wong, M. T. F., (2015). Phosphorus forms in soil solution and leachate of contrasting soil profiles and their implications for P mobility. *Journal of Soils and Sediments, 15*(4), 854–862.

Smith, P., Bustamante, M., Ahammad, H., Clark, H., Dong, H., Elsiddig, E. A., Haberl, H., et al., (2014). Agriculture, forestry and other land use (AFOLU), Climate change 2014: Mitigation of climate change. In: *Contribution of Working Group III to the Fifth Assessment Report of the Intergovernmental Panel on Climate Change*. University Press, Cambridge.

Sohi, S. P., (2012). Carbon storage with benefits. *Science, 338*, 1034–1035.

Syah, R., Zahar, M., & Kianfar, E., (2021). Nanoreactors: Properties, applications, and characterization. *International Journal of Chemical Reactor Engineering, 19*(10), 981–1007.

Tan, X., Wei, W., Xu, C., Meng, Y., Bai, W., Yang, W., & Lin, A., (2020). Manganese-modified biochar for highly efficient sorption of cadmium. *Environmental Science and Pollution Research, 27*, 9126–9134. doi: 10.1007/s11356-019- 07059-w.

Thies, J. E., Rillig, M. C., & Graber, E. R., (2015). Biochar effects on the abundance, activity and diversity of the soil biota. In: Lehmann, J., & Joseph, S., (eds.), *Biochar for Environmental Management: Science, Technology and Implementation* (2nd edn., pp. 327–389). Earthscan Publications Ltd., London, UK.

Tilman, D., Balzer, C., Hill, J., & Befort, B. L., (2011). Global food demand and the sustainable intensification of agriculture. *Biol. Sci., 108*(50), 20260–20264. https://doi.org/10.1073/pnas.1116437108.

Tiwari, D. K., Dasgupta-Schubert, N., Cendejas, L. V., Villegas, J., Montoya, L. C., & García, S. B., (2014). Interfacing carbon nanotubes (CNT) with plants: Enhancement of growth, water and ionic nutrient uptake in maize (*Zea mays*) and implications for nano-agriculture. *Appl. Nanosci., 4*, 577–591.

Ulyett, J., Sakrabani, R., Kibblewhite, M., & Hann, M., (2014). Impact of biochar addition on water retention, nitrification carbon dioxide evolution two sandy loam soils. *European Journal of Soil Science, 65*(1), 96–104.

Utomo, W. H., Islami, T., Wisnubroto, E., & Soelistyari, T., (2017). Biochar as a carrier for nitrogen plant nutrition: 3. effect of enriched biochar on rice (*Oryza sativa* L.) yield and soil qualities. International *Journal of Applied Engineering Research, 12*, 10426 –10432.

Van, D. K. C., Lesschen, J. P., & Oenema, O., (2016). Phosphorus flows and balances of the European union member states. *Sci. Total Environ., 542*, 1078–1093.

Wali, F., Naveed, M., Bashir, M. A., Asif, M., Ahmad, Z., Alkahtani, J., Alwahibi, M. S., & Soliman, E. M., (2020). Formulation of biochar-based phosphorus fertilizer and its impact on both soil properties and chickpea growth performance. *Sustainability, 12*, 9528. doi: 10.3390/ su12229528.

Wang, C., Wang, R., & Wang, Z., (2018). Effects of nano-carbon water-retaining fertilizer on yield and nitrogen and phosphorus utilization efficiency of tuber mustard. *Asian Agric. Res., 10*, 62–65.

Wang, S., Dai, G., Yang, H., & Luo, Z., (2017). Lignocellulosic biomass pyrolysis mechanism: A state of- the-art review. *Prog. Energy Combust. Sci., 62*, 33–86.

Wang, Y., Han, Z., Zhang, Z., et al., (2016). Biochar application promote growth parameters of soybean and reduces the growth difference. *Communications in Soil Science and Plant Analysis, 47*(12), 1493–1502.

Wang, Z. Y., Chen, L., Sun, F. L., Luo, X. X., Wang, H. F., Liu, G. C., Xu, Z. H., et al., (2017). Effects of adding biochar on the properties and nitrogen bioavailability of an acidic soil. *Eur. J. Soil Sci., 68*, 559–572.

Wen, P., Wu, Z., Han, Y., Cravotto, G., Wang, J., & Ye, B. C., (2017). Microwave-assisted synthesis of a novel biochar-based slow-release nitrogen fertilizer with enhanced water retention capacity. *ACS Sustainable Chemistry & Engineering, 5*, 7374–7382. doi: 10.1021/ acssuschemeng.7b01721.

Wen, P., Wu, Z., Han, Y., Cravotto, G., Wang, J., & Ye, B. C., (2017). Microwave-assisted synthesis of a novel biochar-based slow-release nitrogen fertilizer with enhanced water-retention capacity. *ACS Sustain. Chem. Eng., 5*(8), 7374–7382.

Wrobel-Tobiszewska, A., Boersma, M., Adams, P., Singh, B., Franks, S., & Sargison, J. E., (2016). Biochar for eucalyptus forestry plantations. *Acta Hortic., 1108*, 55–62.

Wu, J., Li, Z., Huang, D., et al., (2020). A novel calcium-based magnetic biochar is effective in stabilization of arsenic and cadmium co-contamination in aerobic soils. *J. Hazard Mater., 387*, 122010.

Xiu, S., Shahbazi, A., & Li, R., (2017). Characterization, modification, and application of biochar for energy storage and catalysis: A review. *Trends Renew. Energy, 3*, 86–101.

Xu, Y., Seshadri, B., Sarkar, B., Wang, H., Rumpel, C., Sparks, D., Farrell, M., Hall, T., Yang, X., & Bolan, N., (2018). Biochar modulates heavy metal toxicity and improves microbial carbon use efficiency in soil. *Sci. Total Environ., 621*, 148–159.

Yang, Y., Zhou, B., Hu, Z., & Lin, H., (2020). The effects of nano-biochar on maize growth in northern Shaanxi province on the loess plateau. *Applied Ecology and Environmental Research, 18*(2), 2863–2877.

Yeboah, S., Zhang, R., Cai, L., Li, L., Xie, J., Luo, Z., Wu, J., & Antille, D. L., (2017). Soil water content and photosynthetic capacity of spring wheat as affected by soil application of nitrogen-enriched biochar in a semiarid environment. *Photosynthetica, 55*, 532–542. doi: 10.1007/s11099-016-0672-1.

Yuan, Y., Bolan, N., Prévoteau, A., Vithanage, M., Biswas, J. K., Ok, Y. S., & Wang, H., (2017). Applications of biochar in redox-mediated reactions. *Bioresour. Technol., 246*, 271–281.

Zhang, K., Chen, L., Li, Y., Brookes, P. C., Xu, J. M., & Luo, Y., (2016a). The effects of combinations of biochar, lime, and organic fertilizer on nitrification and nitrifiers. *Biol. Fertil. Soils, 53*, 77–87.

Zhang, P., Huang, P., Xu, X., Sun, H., Jiang, B., & Liao, Y., (2020). Spectroscopic and molecular characterization of biochar-derived dissolved organic matter and the associations with soil microbial responses. *Sci. Total Environ., 708*, 134619.

Zhao, Q., Ma, C., White, J. C., et al., (2017). Quantitative evaluation of multi-wall carbon nanotube uptake by terrestrial plants. *Carbon, 114*, 661–670.

Zhou, B. B., Chen, X. P., Wang, Q. J., et al., (2018). Effects of nano carbon on soil erosion and nutrient loss in a semi-arid loess region of Northwestern China. *International Journal of Agricultural Biological Engineering, 11*(1), 138–145.

CHAPTER 7

Role of Nanoparticles and Nano-Biochar in Reducing Salinity Stress for Improving Soil Health and Sustainable Crop Production

PALLAVI SAXENA,[1] RASHI MIGLANI,[2] DIKSHA SHAH,[2] VISHNU D. RAJPUT,[1] and TATIANA MINKINA[1]

[1]*Academy of Biology and Biotechnology, Southern Federal University, Russia*

[2]*Department of Zoology, D.S.B. Campus, Kumaun University, Nainital, Uttarakhand, India*

ABSTRACT

In the era of climate change, land salinization is becoming a major hazard to soil conditions, leading to a significant reduction in crop productivity. It is estimated that approximately 10 million hectares of land per year are becoming unsuitable for agricultural practices due to salinity. Salinity not only affects plant development and efficiency by inducing oxidative stress, disrupting hormonal balance, and causing nutrient imbalances, but it also has adverse effects on soil properties such as organic matter and C:N ratio. Additionally, high soil salt concentration reduces microbial activity and alters microbial communities in the soil.

Nanoparticles and nano-biochar, with their specific physical and chemical properties such as high surface area, strong zeta potential, and larger pore volumes, have the potential to improve soil conditions and alleviate salinity stress, thereby promoting crop production. The application of nanoparticles

Nanomaterials and Nano-Biochar in Reducing Soil Stress: An Integrated Approach to Sustainable Agriculture. Vishnu D. Rajput, & Tatiana Minkina (Eds.)

and nano-biochar enhances soil cation exchange capacity, organic matter content, and surface area under salinity stress. Furthermore, it helps limit the uptake of Na^+ and promotes the uptake of other essential minerals in saline soil, which are crucial for enhancing plant growth. The application of nanoparticles and nano-biochar also improves the regulation of stomatal conductance and phytohormones. These advancements can be beneficial in increasing crop productivity to meet the demands of a growing global population and in rejuvenating agricultural land for sustainable practices.

7.1 INTRODUCTION

Biochar (BC) is a carbon-rich solid substance produced by the pyrolysis of biomass at high temperatures and in an oxygen-free (or low-oxygen) environment. Biochar can be used in various applications due to its exceptional features, including biocompatibility, simplicity of production, ease of regeneration, and tunable surface. Biochar has been widely studied for the removal of harmful pollutants, where it is used to immobilize harmful substances such as heavy metals and organic pollutants through adsorption and as a support for various types of reagents that can be used to degrade hazardous toxins in unconventional oxidation processes (O'Connor et al., 2018). Furthermore, Biochar production is not only effective and environmentally friendly, but also cost-effective in reusing waste materials, as it requires very little energy and can be produced at temperatures below 700°C (Saxena et al., 2017; Ashiq et al., 2019; Ashiq & Vithanage, 2020). Pyrolysis technology determines the size of biochar particles, ranging from micrometers to centimeters. Further reducing the size of biochar particles to the nanoscale range, up to 100 nm or less, enhances their characteristics and results in a larger surface-to-volume ratio, increased surface energy, adsorption capacity, and consequently, organic efficiency (Naghdi et al., 2019; Hosny et al., 2022). Chen et al. (2011) confirmed that nano biochar with a high hemicellulose concentration exhibited a greater surface area than nano biochar with a high lignin content. This is likely due to the lower resistance of lignin pyrolysis at temperatures below 1,000°C (Neeli & Ramsurn, 2018).

So, nano-biochar is getting more popular because it's like biochar and nanotech combined (Chen et al., 2017). Nano biochar has more surface area, is smaller, and has more oxygen stuff than regular biochar (Pratap et al., 2022). People are using nano biochar for soil, sensors, cleaning up pollution, and other stuff (Xioa et al., 2020; Zhang et al., 2022). In soil, nano biochar helps bacteria and fungi grow and keeps bad stuff out. It also helps soil stay healthy and breaks down pollution (Quilliam et al., 2013; Ameloot et al., 2015; Jiang et al., 2023). Things like heat, water, and pollution can hurt farming (Badawy et al., 2021). Abiotic stress affects plant morphology, physiology, and biochemistry depending on crop type, stress type, and duration of exposure (Semida et al., 2014; Rady et al., 2021; Shaaban et al., 2022). Salinization is a prevalent abiotic environmental stress that substantially reduces plant development (Taha et al., 2021; Khan et al., 2021). Salinization is a global agricultural issue, particularly in dry and semi-dry countries. High salt levels impair soil and plant physiology (Rajput et al., 2015). Salinity stress leads to nutrient imbalance, membrane damage, impairment in detoxifying reactive oxygen species (ROS), and reduced photosynthetic activity (El-Saadony et al., 2022). According to estimates, 15% of the globe's total land area has been damaged by soil erosion and salinization, which subsequently degrades the soil's biological and physicochemical characteristics (Tóth et al., 2008; Yadav et al., 2019).

Nanotechnology improves crop output and quality, farming, and world food demand. Many NPs may reduce abiotic stress-induced plant damage and increase crop output. Together with nanoscience and other environmentally subtle agronomic technologies, species and varieties designed to handle these trials could be the most effective variation approach to cope with environmental change and abiotic stress aspects, which would lead to sustainable production (Kumari et al., 2022; El-Saadony et al., 2022). In particular, nanomaterials provide an effective solution for degrading perilous contaminants, while the application of nano-biochar plays a pivotal function in increasing plant progression, reducing carbon sequestration, and global warming (Chausali et al., 2021).

7.2 IMPACT OF SALINITY ON DIFFERENT PROPERTIES OF SOILS

The global soils are intimidated by handful ecological issues triggered by climate amendment, population growth and economic development (Prăvălie et al., 2021; Rajput et al., 2023). Salinity is a prevalent abiotic stress that is accountable for a substantial reduction in plant growth. High levels of Na^+, Mg^{+2}, K^+, Ca^{+2}, Cl^-, HCO_3^{2-}, SO_4^{-2} and B ions in soil are the primary contributor to its salinity, and these ions have been shown to inhibit the growth of plants (Khaled & Fawy, 2011; Yadav et al., 2019). As a consequence, the high salt content decreases the osmotic potential, i.e., elevated Na^+ absorption in the soil suspension, depresses nutrient activity and generates an utmost total of Na^+/Ca^{2+} or Na^+/K^+, which hinders water uptake and slows down seed growth (Yadav et al., 2019; Seleiman et al., 2020; Taha et al., 2021). A saline topsoil is typically characterized by an electrical conductivity (EC) of the saturation extract (EC_e) in the root zone that exceeds 4 dS m^{-1} (about 40 mM NaCl) at 25°C and a 15% exchangeable salt concentration (Table 7.1). At this EC_e, the yield of most crop plants is diminished, while several crops also experience yield decrease at lower EC_{es} (Jamil et al., 2011). Over 2,000 acres of arable land are abandoned each day due to salinization on a global scale, contributing 1 to 2% of the world's agricultural soil each year (Zaman et al., 2018). Excessive soil salinity is hazardous not only to the chemical and physical qualities of the soil, but also to plants, microbes, and other soil-dwelling creatures (Leogrande & Vitti, 2019).

On one side, the extreme salt content in the soil solution separates Na ions and expands clay particles, causing soil development and distribution, weakening soil structure, and decreasing the penetrability coefficient, water intrusion, and porosity (Amini et al., 2016). An increase in salinity has negative consequences on biologically mediated soil activities, such as soil respiration (Ghollarata & Raiesi, 2007). Apart from this, high salinity causes elevated pH, damage to organic matter, reduction in nutrient cycling (both C and N mineralization), nutrient inadequacy, and toxicity of selected substantial anions and cations in the soil (Patel et al., 2017; Dagar et al., 2019). In general, salt-affected soils display nutritional deficits (mostly nitrogen, phosphorus, and potassium) and reduced accessibility of micronutrients (such as aluminum, iron, zinc, copper, and manganese), probably due to elevated pH values (Fageria et al., 2011; Rani et al., 2019).

In high salt conditions, microbial communities tend to shift towards lower metabolic functioning to adapt salt-tolerant mechanisms (Yuan et al., 2007). Additionally, the metabolic quotient is found to increase under saline conditions, reflecting the stress on microbial societies (Wong et al., 2008). Moreover, it has been observed that C and N mineralization rates are drastically reduced at high salt concentrations. The increased rate of nitrification accelerates volatilization and subsequently leads to the loss of ammonia in the soil. Soil enzymes such as urease (for nitrogen mineralization), phosphatase (for phosphorus mineralization), and β-glucosidase (for carbon mineralization) play a significant role in nutrient cycling (Dagar et al., 2019; Davies et al., 2022). Many investigations have confirmed that increasing soil salinity inhibits a wide variety of soil enzymatic activities. These include β-glucosidase, benzoyl-argininamide, alkaline phosphatase, dehydrogenase,

TABLE 7.1 Types of Salt Affected Soils

	pH	**Salt Concen-tration**	**Electrical Conductivity**	**Exchangeable sodium percentage**	**Sodium adsorption ratio**	**Most important Cation**	**Most important Anion**	**Intrusion**	**Drainage**
Saline (Alkaline)	Below 8.5	Above 0.2 %	Above 4.0 dSm^{-1}	Below 15.0%	Below 13.0	Ca^{2+}, Mg^{2+}, K^{+}	Cl^{-}, SO_4^{2-}, NO_3^{-}	Significant	Significant
Sodic (Alkali)	Above 8.5	Below 0.2 %	Above 4.0 dSm^{-1}	Above 15.0%	Above 15.0	Na^{+}	CO_3^{2-}, HCO_3^{-}	Not Significant	Not significant
Saline – Sodic	Above 8.5	Above 0.2 %	Above 4.0 dSm^{-1}	Above 15.0%	Above 15.0	Ca^{2+}, Mg^{2+}, K^{+}, Na^{+}	Cl^{-}, SO_4^{2-}, NO_3^{-}, CO_3^{2-}, HCO_3^{-}	significant	Significant

invertase, urease, amylase, and catalase (CAT) (Tiwari et al., 2019; Dagar et al., 2019). The soil's enzymatic activities are responsible for most of the biochemical changes that occur in the soil, and these variations in the soil characterize the soil quality (Sahab et al., 2021).

7.3 PHYSICO-CHEMICAL PROPERTIES OF NANO BIOCHAR

Heating feedstock to temperatures between 350 to 700°C in an oxygen-deficient environment with less than 1% oxygen in a process called pyrolysis generally yields biochar (Song et al., 2019). By employing an assortment of thermochemical procedures, such as gradual and rapid pyrolysis, torrefaction, carbonization, i.e., flash/hydrothermal, and gasification procedures, biochar is synthesized from biomass rich in lignocellulosic materials, such as agricultural and timber residues, animal litter, etc. (Li et al., 2017). While the properties of nano-biochar (N-BC), which is produced by shrinking the size of the same biochar's particles, may resemble those of macro-biochar, it frequently exhibits properties that are very dissimilar as those at the macro measure, such as surface area. Moreover, N-BCs are selected for a variety of relevance in part mainly because of their inherent characteristics. However, the use of nano-biochar can also be anticipated based on feedstock parameters. For instance, N-BC derived from plants through a relatively bigger aromatic cluster magnitude (hydrophobic segmentation of organic impurities) along with an additional oxygen-comprising surface functional group may have a strong kinship for organic contaminants as well as the ability to remove contaminants by binding (Khare, 2021). Nevertheless, N-BC generated from municipal waste materials that are rich in alumino-silicate and carbonate sulfate can help with the co-precipitation or chelation of heavy metals (Song et al., 2019). Like this, N-BC efficacy as an adsorbent and catalytic agent can depend on the degree and type of apparent functionality and pore configuration.

7.4 PHYSICAL PROPERTIES

7.4.1 SURFACE AREA

Presence of a substantial number of pores confers a high surface area to the biochar (Brewer et al., 2014). Nano biochar (N-BC) has a surface area that varies greatly and lies in the range between 0.76 to 1,736 $m^2\ g^{-1}$ (Khare,

2021). Depending on the feedstock and fabrication techniques utilized for N-BC, it generally achieves a higher surface area in comparison to the bulk biochar (B-BC). According to Zhang et al. (2019), N-BC has a surface area increase of 8–9 $m^2 g^{-1}$ when associated with B–BC. In accordance with the assembly processes, the surface area varied between 9.08 and 173 $m^2 g^{-1}$ for carbonization, 3.67 and 1,736 $m^2 g^{-1}$ for ball milling, 21.7 and 253 $m^2 g^{-1}$ for centrifugation, and 0.76 and 264 $m^2 g^{-1}$ for sonication. At various pyrolysis temperatures, however, there have been noticeable variations in the external area. Nanobiochars generated at low temperatures, 300–400°C, possess less superficial areas, ranging from 5.6 to 47.2 $m^2 g^{-1}$, correspondingly, but high temperatures of 450 and 600°C produce nanobiochars with a wider surface area reaching between 342 and 430 $m^2 g^{-1}$ (Huang et al., 2020; Xu et al., 2019; Li et al., 2020). It is predicted that the boost in area would result from the release of gases caused by the devolatilization of biomass resources and the formation of pores on the material's surface (Oleszczuk et al., 2016).

Moreover, the high surface area could be explained by the pyrolysis-induced demolition of alkyl and ester functional groups. The treatment of ball milling on B-BC to produce N-BC results in a larger external area than any other method employed to prepare N-BC. This might be ascribed to either an increase in external surface area due to a reduction in grain size or an expansion in interior surface area owing to opened inner pore networks (Lyu et al., 2018; Wang et al., 2018). On the other hand, sonication-based exfoliation might not improve the surface functionality of N-BC. According to Song et al. (2019), aliphatic chains and the -COOH, -OH, and -C– –H functional groups are more abundant in N-BC than in B-BC. Nevertheless, Naghdi et al. (2017) demonstrated that N-BC integrated using ball milling did not differ significantly from B-BC in terms of surface functioning. In addition to this, N-BC had greater O/C ratios than B-BC, indicating that the surface of N-BC contains more O-containing functional groups (Liu et al., 2018). Moreover, Ramezanzadeh et al. (2021) noted that the balling procedure transformed longitudinal pores and canals into flat pieces, thereby increasing surface area and overall pore volume. Nevertheless, N-BC made from mineral rich rice husk had a smaller surface area, i.e., 298 $m^2 g^{-1}$ compared to that made from more cellulose rich corn straw, which had a larger surface area i.e., 364 $m^2 g^{-1}$ (Ma et al., 2019). Like this, N-BC produced using a probe-type ultrasonic vibrator from wheat straw, wicker, and elephant grass (Miscanthus) had specific surface areas of 29.56 ± 1.1, 18.25 ± 1.5, and 36.39 ± 2.9 $m^2 g^{-1}$, respectively (Wang et al., 2018). The amount of hemicellulose in the biomass, however, determines the surface area. Nanobiochar made from lignin biomass exhibits a relatively reduced surface area compared to that made from high hemicellulose biomass, even under elevated temperature conditions (Chen et al., 2011).

7.4.2 FIXED CARBON, VOLATILE MATTER, AND YIELD

Comparing N-BC made from B-BC at high temperature and low temperature, the former has a higher bulk density, carbon content, in addition to extractable cations like P, Ca, Mg, Fe, K, Zn, and Mn (Nath et al., 2019). Likewise, the carbon contents of N-BC change depending on the raw material employed for synthesis. For instance, N-BC derived from coconut shells (90–94%) had more carbon than N-BC derived from sewage sludge (4%). Also, according to recent research, the N-BC differs from the corresponding B-BCs in having a slightly lower carbon concentration and slightly higher ash content. Moreover, compared to the B-BC, the N-BC contained less aromatic and carbonized carbon (Wang et al., 2013).

Nevertheless, biomass releases volatile compounds at temperatures of about 950°C (Basu, 2013). An approximate analysis (wt%) of a softwood nanobiochar showed that it contained 71.3% fixed carbon and 25% volatile materials (Wallace et al., 2019). Also, the amount of volatile matter released depends on the type of biomass, and some softwood biomasses, like cinnamon, might release supplementary volatile matter than other softwood biomasses. It is interesting to note that pinewood nanobiochar, which is made from resinous hardwood, has been shown to have 97% volatile matter and only 1% fixed carbon (Naghdi et al., 2019). About the yield of N-BC, there is very little information in the available literature. N-BC synthesized from two feedstocks, namely soybean straw and cattle dung, were compared in terms of yield by Guo et al. (2020) and reported that the former had a higher N-BC output.

According to Liu et al. (2018); and Nath et al. (2019), as the pyrolysis temperature of peanut shell B-BC increased from 300°C to 600°C, the N-BC yield gradually decreased from 1.41% to 2.36% to 0.47% to 0.87%, respectively. It might be associated with the breakdown of a more stable graphitic and less stable amorphous component of biochar. In addition, a positive correlation involving the color and the yield of nano-biochar was noted, i.e., higher yields were obtained from suspensions with darker colors than from suspensions with lighter colors (Song et al., 2019). For example, barley grass biochar suspensions with dark color yielded 20.5%, while pinewood biochar suspensions with light color yielded 1.43% (Song et al., 2019).

7.4.3 WATER-HOLDING CAPACITY

The hydrologic aspects of biochar are directly influenced by the pyrolysis procedure and the parent material's porosity. The number of surface functional groups decreases as pyrolysis temperatures rise, consequently reducing the water binding capacity and increasing hydrophobicity (Alam et al., 2018a & b). Hence, the affinity towards water increases as the number of functional groups increases (Weber & Quicker, 2018). As a result, based on pyrolysis temperature, there is no noticeable variation in water-holding capacity as particle size decreases. The porosity and arrangement of the pores also play a role in a particle's capacity to hold water. According to Naghdi et al. (2017), N-BC had a homogeneous tiny pore with a standard pore size of 16.3 Å, as opposed to the typical pore size of 31.6 Å in the case of B-BC. However, pinewood nanobiochars reportedly exhibit water retention capacities of 9.75 and 2.7 Å, respectively, when produced by pyrolysis at 525 and 400°C (Naghdi et al., 2019; Yu et al., 2013). Because of inadequate interconnectivity and coagulation of the pores within residual factors, biochar with minimal assembly temperatures has pores that are too small to retain a significant quantity of water (Das & Sarmah, 2015). Surprisingly, heating dramatically expanded the pore volume of altered N-BC, however, it did not notably change the pore dimension (Jenie et al., 2017).

7.4.4 MECHANICAL STABILITY AND GRINDABILITY

The mechanical stability of biochar is inversely associated with its porosity (Kumar et al., 1999). As highly porous biochars are weak, processing techniques like grinding lead to structural fissures that allow for the liberation of volatile substances and the evaporation of water. As a result, nanobiochar is often weak in strength (Noumi et al., 2014). A better comparison of mechanical stability, however, is difficult to develop because of the relatively small amount of research. Given the relationship between mechanical stability and porosity, nanoscale particles will exhibit greater mechanical stability than macrobiochars due to their lower porosity. Torrefaction, often referred to as modest pyrolysis, adds brittleness to the biomass during the formulation procedure, which enhances grindability (Weber & Quicker et al., 2018). Biomass that has been torrefied and contains a significant amount of hemicellulose has adjusted grindability, permitting the use of the Hardgrove grindability index (HGI) (Phanphanich & Mani, 2011). Prior to torrefaction, the fiber analysis of the biomass would yield information that would be helpful in selecting biomass with high levels of hemicellulose (Chen et al., 2011). Consequently, the grindability of biochars is governed by both the preparation methods and the feedstock components. The interior tissues' chemical and physical structures, such as cellulose, hemicellulose, and lignin, are destroyed by torrefaction (Zhang & Zhang, 2019). The ideal settings for optimal grindability of bio coal, according to Chen et al. (2011), are 24,000 revolutions per minute (rpm) for 1 minute. Therefore, it is possible to evaluate various milling procedures for reducing particle size based on the energy needed for grinding, which is dependent on the production temperature.

7.4.5 ZETA POTENTIAL AND COLLOIDAL STABILITY

The zeta potential has been evaluated to interpret the particles' aggregation, electrokinetics, and solidity. It further reflects the surface charge of the particles (Safari et al., 2019; Song et al., 2019). When surface functional groups, which normally include oxygen, are deprotonated, nanobiochars have a noticeable negative surface charge (Yi et al., 2015). Moreover, the feedstock's type and particle size have an impact on biochar's zeta potential. Hence, in contrast to comparable micro or macrobiochars, nanobiochars have lower stabilities. Additionally, biochar with a large zeta potential showed less aggregation and more dispersion of the particles. The zeta potentials of N-BCs ranged from -19.4 to -87 mV and were relatively higher than those of B-BC. According to research by Oleszczuk et al. (2016), biochar-NPs have a greater zeta potential than the original materials (wicker, wheat straw, and miscanthus) by a factor of 56 to 152%. Nevertheless, Huang et al. (2017) showed that the zeta potential of N-BC derived from B-BC of bamboo dust increased by 17%. These results indicated high dispersity and colloidal stability of N-BC. Zeta potential comparisons revealed that the N-BC produced from wicker and wheat straw biochar had greater colloidal strength than that produced via sewage sludge biochar. Moreover, a higher zeta potential was detected in the N-BC synthesized using centrifugation techniques.

7.4.6 AGGLOMERATION POTENTIAL

Units agglomerate into pellets or granules with improved strength and density. Accumulation that occurs during tumbling is often referred to as growth agglomeration, whereas agglomeration that occurs under pressure is known as press agglomeration (Weber & Quicker, 2018). Due to the abundance of carboxylic and phenolic groups, the biomass's lignin content has a substantial impact on the pelleting process. Lignin softens and serves as a natural binder throughout the process. As a result, pellets made from woody biomass with high lignin levels are more stable than those made from low-lignin biomass (Föhr et al., 2017). In the temperature range between 90°C (wet-based) and 200°C (dry-based), the lignin softening point is reportedly influenced by the moisture content of the biomass (Thek et al., 2012). In addition to the lignin content, the stability of pellets is further improved by particle rearrangement and by filling intra-particle gaps (Weber & Quicker, 2018). Agglomeration potential, however, is reportedly dependent on the biomass's composition (Song et al., 2019). Moreover, under acidic conditions, protonation of functional groups and the subsequent drop in surface charge may cause nano biochar particles to start to aggregate (Ma et al., 2019).

7.4.7 FLUORESCENCE ACTIVITY

Carbon dots made from various carboniferous feedstocks and intended for heavy metal sensing have been detected and reported to haveYP have the fluorescence property; however, investigations on nanobiochars are missing. In a recent work, three distinct nanobiochars—derived from rice straw, dairy manure, and sorghum straw—were investigated as potential probes for sensing Pb, Hg, Cu, and Ni using fluorescence recovery and fluorescence quenching (Plácido et al., 2019a). It is hypothesized that the non-shared electron pairs of the carboxylic and phenolic functional assemblies in nanobiochars, which include the C—O, C—O, and C—OH groups, form coordination connections with heavy metal ions (Plácido et al., 2019b). Notably, the same study revealed that nanobiochars were more expensive than pure biochar (0.076 pounds per kilogram) and cost just 1% as much as carbon dots (129 pounds per kilogram) to produce as electrode sensor materials. The fluorescence characteristics of nanobiochar and their application in the creation of sensors, however, require more in-depth research because some of the metals demonstrated lower efficacy.

7.5 CHEMICAL PROPERTIES

7.5.1 ELEMENTAL COMPOSITION

Synthesizing biochar changes the chemical makeup of raw material predominantly into carbon. An obvious decline in the carbon substance has been noticed with declining particle size, from the macro to nano scale. For instance, the carbon concentrations of wood chip biochar at the macro, micro, and nano sizes were reported to be 81.2%, 63%, and 17.7%, respectively (Song et al., 2019). Oxygen and hydrogen levels are decreased because of the separation of functional groups, while the carbon content is proportionally raised (Weber & Quicker, 2018). Moreover, the aromaticity and stability of biochar are indicated by its H:C ratio (Zhang & Ok, 2014). Hemp and softwood nano biochars exhibit increased carbon content in the H:C ratio along with graphitic-identical structures, which could improve the intensity composite material's structure (Wallace et al., 2019). Likewise, the H:C ratio of nano-biochar made from pinewood has been found to be 0.5 (Naghdi et al., 2019). Moreover, Elephant grass (Miscanthus), wicker, and corn straw nanobiochars all had an H:C ratio of 0.5, but wheat straw nano biochar had an H:C ratio of 0.66 (Oleszczuk et al., 2016; Ma et al., 2019). However, of all the nanobiochars, nano biochar made from rice husk has the lowest reported H:C ratio of 0.17 (Ma et al., 2019). However, when compared to nanoscale biochar, the H:C ratio of macrobiochars made from identical feedstock materials does not show any appreciable variation. Although wheat straw biochar was produced by Oleszczuk et al. (2016); and Zhang et al. (2019a) at a similar pyrolysis temperature, the H:C ratios were not the same (0.66 and 0.02, respectively), which might be because both nano and colloidal fractions were included in the resulting ball-milled biochar (Zhang et al., 2019b). The significant differences in characteristics between ball-milled biochar and nano biochar, which contains both colloidal and nano-sized fractions, are confirmed by the superior H:C ratio in nano biochar. Furthermore, the O:C ratio of a biochar exposes its polarity. Elephant grass and wicker nano biochar have an O:C ratio of 0.21, compared to 0.48 for wheat straw nano biochar. According to a relatively recent study, both H:C and O:C ratios tend to increase as particle size decreases (Song et al., 2019).

7.5.2 STRUCTURAL COMPOSITION

Plant-based biomass is abundant in cellulose, hemicellulose, and lignin, all of which contribute to the strength of the formation. Due to the complete and partial degradation of hemicellulose, cellulose, and lignin biomolecules at temperatures between 250 and 350°C during pyrolysis, most of the characteristics of biomass undergo significant changes (Ramanayaka et al., 2020a). The Fourier-transform infrared (FTIR) bands of softwood nano biochar were reported by Wallace et al. (2019), indicating the presence of hemicellulose, cellulose, and lignin. In microwave pyrolysis, the pore geometries of softwood and hemp nano biochar show signs of cracking despite the presence of these chemical components. The pore geometries of hemp and softwood nano biochar show signs of cracking with increasing power levels in microwave pyrolysis, regardless of the presence of these chemical components. Additionally, the mineral composition of biochar is strongly influenced by the feedstock resources. Plant-based biochars, such as wood and agricultural waste, show a dominant SiO_2 peak, whereas sludge, municipal, and animal-sourced biochars show aluminosilicate, $Ca_3(PO_4)$, and $CaCO_3$ mineral peaks. In contrast to macro biochars, nano biochars also showed strong mineral peaks (Song et al., 2019; Ramanayaka et al., 2020b).

7.5.3 pH

The preparation process alters the pH of the final biochar significantly among the pyrolysis technique and hydrothermal carbonization (Weber & Quicker, 2018). The alkalinity of the biochar, which may be acidulated and have its pH lowered by an acid-activation phase, is also significantly impacted by the kind of feedstock. For agricultural uses, the pH of biochar is crucial since it is utilized as a soil supplement. Pinewood nano biochar, a high-cellulose and lignin-containing biochar, exhibits a pH of 6.6 (Naghdi et al., 2019), which is mildly alkaline. Yet, while having relatively little quantities of cellulose and lignin, elephant grass (Miscanthus), wicker, and wheat straw nano biochar have very basic pH values of 7.8, 8.9, and 9.1, respectively (Oleszczuk et al., 2016). Carbonates, soluble organic compounds, surface organic functional groups, and other inorganic alkalis are the four main groups that have an impact on the basic nature of biochar. Depending on the feedstock material, the presence of organic substances like cellulose and lignin, as well as essential oils and resins, can have a direct impact on a biochar's alkalinity. Similar pH values (6.4 and 6.8) are displayed by goethite-modified peanut shell nano biochar at temperatures of 300 and 400°C, respectively. At higher temperatures of 500 and 600°C, in that order, a modest rise in pH (7.08 and 7.28) can be seen; this is due to the elevated ash content created throughout the pyrolysis of biomass (Ahmad et al., 2014; Liu et al., 2018). There are many kinds of biochar; however, those made from wood have high alkalinities (pH > 8). Biochars made from pinewood (pH 6.3) and wood chips (pH 7.05) exhibit weakly acidic or neutral pH, which may be caused by the low ash concentration and carboxyl group loss. Yet, the pH of wood-derived biochars at the nano and microscopic scales is higher (pH 8.9–9.7) than that of the macroscale particles (Fidel et al., 2017). Biochars made from agronomic waste and manure also have superior pH levels (pH 10.0–10.4). The loss of inorganic elements like Ca, P, and S may be the reason why the pH values of the micro and nano fractions are somewhat lower, up to 9.5 (Zhao et al., 2016). The pH values of sewage sludge biochar at the nano, macro, and micro sizes are the same (Song et al., 2019).

7.5.4 SURFACE CHEMISTRY

The feedstock raw material and pyrolysis temperature largely affect the surface functional groups found in nano biochar. The material's volatile organic molecules may devolatilize as the temperature rises, which results in a loss of surface functional groups. Nevertheless, research has shown that compared to pure unmilled biochar (0.8–2.9 mmol g^{-1}), ball-milled biochar (2.2–4.4 mmol g^{-1}) has additional oxygen-containing functional groups (Lyu et al., 2019). Moreover, following ball-milling, there are more acidic surface functional groups, which improves adsorption effectiveness (Lyu et al., 2018; Wang et al., 2020). Additionally, milling shows the nano biochar's graphitic character and causes the generation of ROS, which may enhance the surface chemistry of the material relative to pure biochar. Most of the research has employed FTIR spectroscopy to pinpoint surface functional groups, and some have even used Boehm titration to quantify those groups (Wallace et al., 2019; Lyu et al., 2018). During the milling procedure, it was learned that the concentrations of acidic functional groups in bagasse biochar increased from 0.8 to 2.5 mmol g^{-1} (Lyu et al., 2018).

Thus, this increase may be due to solvents that are added during milling. The quantity of acidic surface functional groups (O—H) can be enhanced by using different solvents, including water and ethanol (Munkhbayar et al., 2013; Ramanayaka et al., 2020b). The most prevalent surface functional groups found on nano biochar are aromatic C=C/C=O, O—H, COOH, and C—H (Qin et al., 2019). Depending on the feedstock material, SiO_2 and N—H bonds are also present in addition to those (Wallace et al., 2019). In comparison to macro biochar, the nano and micro fractions of biochar have been found to contain a relatively high number of functional groups, such as -COOH and COOC, and newly evolving aliphatic chains (Qu et al., 2016; Zhao et al., 2013). This is supported by significant increases in -OH stretching (3,400 cm^{-1}) and C-O stretching in nano biochar samples (Keiluweit et al.,

2010). Peak shifts of -C=C and -C=O with decreasing particle size show the decrease of conjugated benzene rings, and the nano and micro portions of biochar exhibit minimal aromaticity. Responsive oxygen, which is abundant in nano biochar.

7.6 USE OF NANOPARTICLES AND NANO-BIOCHAR IN REDUCING SALINITY STRESS FOR IMPROVING SOIL HEALTH AND CROP PRODUCTIVITY

Nanoparticles are known to have a multitude of different characteristics because of their optimal size bounds and enhanced surface area, making them a popular choice as environmental remediation agents (Chauhan et al., 2021; Sharma et al., 2022). Nanomaterials can effectively eliminate both heavy metals and organic compounds from contaminated soils using nano-zero valent iron (nZVI) (Qian et al., 2020). Nano-zero valent iron (nZVI) particles can convert harmful pollutants into less dangerous compounds, such as converting chromium (VI) to chromium (III) and generating ferrous chromite (Rabbani et al., 2016). Moreover, when biochar is added to nano-zero valent iron nanoparticles (nZVI NPs), it increases the reduction response capacity of nZVI and enhances its removal efficiency, while also reducing the agglomeration of iron particles in the soil. For instance, it has been discovered that mixing nZVI nanomaterials with biochar may remove 66% of the chromium (VI) levels present in soil (Qian et al., 2020). Another study highlighted that the removal of Cd (46.66%), Pb (48.88%), and Zn (47.01%) from contaminated soil was improved by the addition of OA-nZVI NMs at 0.4 g kg^{-1} (Cao et al., 2018). Copper, cadmium, nickel, and zinc were all eliminated from contaminated soils after being treated with biochar made from inexpensive raw materials including rice husk, water hyacinth, and black tea waste (Lobzenko et al., 2022; Awasthi et al., 2022). Wang et al. (2014) also demonstrated that the amount of Cr that was taken up by *Brassica napus* and *B. rapa* subsp. *Pekinensis* increased when nZVI was used with biochar. Nanotechnology stimulated soil enzymes to improve nutrient transport and soil fertility (Zhao et al., 2021). The interaction between Nanomaterials and rhizospheric bacteria can increase both plant development and soil healthiness (Khan et al., 2022). The use of industrial goods based on NMs, such as nano fertilizers, which have been coated with NMs (Verma et al., 2022), exhibited a positive outcome on the soil microbial community (Chaudhary et al., 2021), causing a change in rhizospheric microbiome characteristics and increasing plant growth (Khanna et al., 2021) (Figure 7.1).

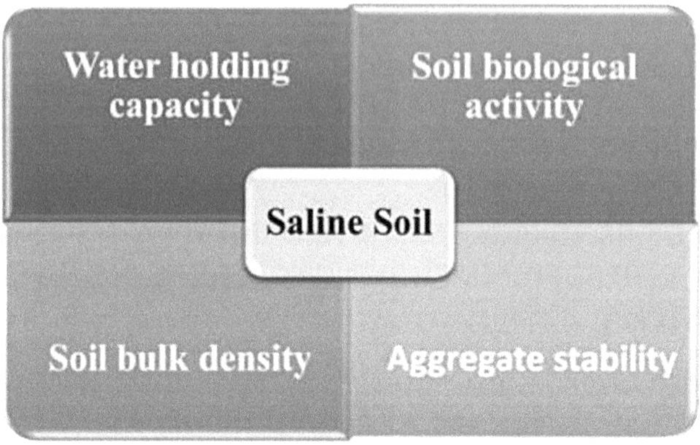

FIGURE 7.1 Soil properties enhanced by the nano-biochar application.

Nano biochar has an expanded surface area and a greater number of functional groups attached to it, which enhances the mechanical and thermal stability of the nano biochar over macro particle biochar (Noreen et al., 2021). Moreover, the reduced hydrodynamic radius, improved zeta potential, carbon defects which may produce reactive oxygen species, resulted in greater adsorption proficiency to the nano-sized biochar compared to the micro one (Ramanayka et al., 2020). Carbon deficiencies are generally occupied by a substantial number of unitary electrons which help in decreasing formation energy, hasten electron transfer, and play a vital role in catalytic activities (Greco et al., 2019). Another study exhibited that the effect of biochar on pH is primarily determined by the raw material used for the biochar preparation and the pyrolysis conditions (Yadav et al., 2016). However, a decrease in soil pH by 1.4% and an increase in phosphorus, iron, and zinc up to 36%, 29%, and 36%, respectively, were evident. Ions like H^+ and Al^{3+} ions probably get exchanged from surfaces and enter the saline soil, subsequently lowering the soil pH. Owing to these properties, nanobiochar is found to have magnificent efficacy in lowering the sodium concentration within the soil due to the absorption capacity of their particles. Hence, nowadays they are considered as a promising bioremediation agent for salinity. Additionally, using nano biochar not only decreases the salt concentration within the soil but also decreases the soil electrical conductivity as well. As salinity and electrical conductivity are dependent and proportional to each other (Chao-Yin et al., 2011). Furthermore, due to salinity, root water uptake is reduced due to the presence of extreme salt conditions, which decreases the osmotic potential of the soil and inhibits water uptake (Prakash & Singh, 2020). The addition of nano-biochar lowers the salt concentration and improves the osmotic potential, enhancing the water retention capacity of the soil (Akhtar et al., 2015) (Figure 7.2).

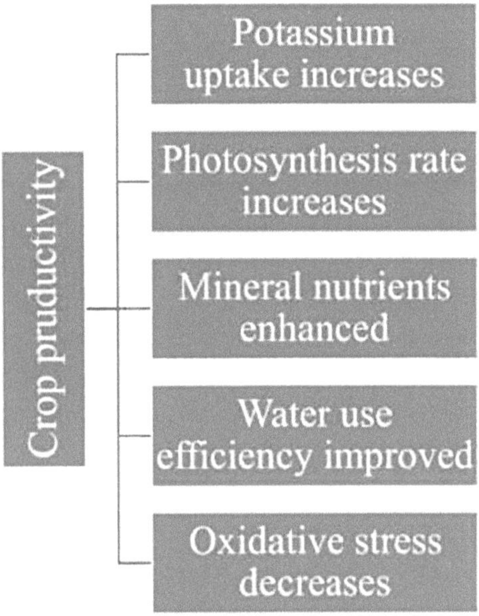

FIGURE 7.2 Nano-biochar mitigate salinity stress by regulating these above-mentioned activities in the plants to achieve high crop productivity.

Several studies supported the improvement of soil quality and crop yield after the usage of nano-biochar. Recently, Khader et al. (2023) reported that the dry and fresh weights of tomato plants increased from 60.23 and 12.4 g in the control to 88.82 and 24.65 g in treatments after the application of nano-biochar. Moreover, the number of flowers and the length of roots were also found to be positively affected. Similarly, Yang et al. (2020) also reported that the treatment of biochar in alternate root-zone drying irrigation drastically enhanced plant height, shoot biomass, and grain by 11.7%, 18.8%, and 10.2% under salinity in Quinoa. The study reported that the synergy between biochar and the root zone efficiently modified the equilibrium concerning chemical signals and hydraulic signals. This helps increase the intrinsic water use efficiency and crop yield by mitigating the salinity impact. Similarly, Rashid et al. (2023) studied nano-biochar along with copper oxide nanoparticles, which improved the quality of clayey loam soil and increased copper uptake and crop productivity in wheat crops. The study reported that wheat biological, grain yield, and nitrogen uptake were enhanced by 35%,

62%, and 80% in the treated wheat crop plants. Furthermore, a recently conducted study showed that after the application of simple nano-biochar, there were elevated levels of nitrogen and potassium up to 46% and 48% in salinity-affected soil (Moradi et al., 2019). Moreover, when biochar was applied along with rock phosphate and cow manure, improvement in soil nutrient status was observed significantly. As nano biochar has the high specific surface area which leads to the elevated cation exchange capacity. It may elevate processes like denitrification and sublimation, due to which total nitrogen within the soil was found to increase after biochar application (Clough et al., 2013). A substantial increase in potassium content was also evident after the biochar application due to the presence of ash, sodium, magnesium, and calcium content in high amounts (Lin et al., 2015). Khan et al. (2021) also testified that nano biochar provides high nutrient content for an extended period and could enhance crop productivity. Microbial communities are also stated to be replenished after the application of nano biochar. It is stated that the porous nature of the particle plays a necessary role in it. The characteristic organic surface of nano biochar not only provides sufficient nutrients for the microbial communities but also delivers protected ecological niches (Nguyen et al., 2018). The presence of sufficient nutrients in saline soil also increases bacterial transformation, and thus the status of soil through bacterial conversions improves (Bhaduri et al., 2016).

7.7 CONCLUSION

Nano biochar could play a fundamental role in enriching the nutrient status of soil and subsequently enhance crop yield under saline conditions. Nano biochar, due to its magnificent efficacy for salt adsorption, could easily create harmonious conditions for micro-macro nutrient cycling/uptake. The presence of nutrients will surely rejuvenate the microbial community's status within the soil and help in efficient nutrient translocation within the plants, even in saline conditions. Further research may be needed to use nano biochar in fertilizer form for agricultural purposes. It would be a cost-effective, eco-friendly, and sustainable approach.

ACKNOWLEDGMENT

Authors acknowledge the financial support from the Strategic Academic Leadership Program of the Southern Federal University ("Priority 2030").

KEYWORDS

- **nano-biochar**
- **nanoparticles**
- **salinity**
- **soil health**
- **stress**

REFERENCES

Akhtar, S. S., Andersen, M. N., & Liu, F., (2015a). Biochar mitigates salinity stress in potato. *J. Agron. Crop Sci., 201*, 368–378, doi: 10.1111/jac.12132.

Ameloot, N., Sleutel, S., Das, K. C., Kanagaratnam, J., & De Neve, S., (2015). Biochar amendment to soils with contrasting organic matter level: Effects on N mineralization and biological soil properties. *GCB Bioenergy, 7*, 135–144. https://doi.org/10.1111/gcbb.12119.

Amini, S., Ghadiri, H., Chen, C., & Marschner, P., (2016). Salt-affected soils, reclamation, carbon dynamics, and biochar: A review. *J. Soils Sediments, 16*, 939–953. https://doi.org/10.1007/s11368-015-1293-7251726.

Ashiq, A., & Vithanage, M., (2020). Biochar-mediated soils for efficient use of agrochemicals. In: Narasimha, M., & Prasad, V., (eds.), *Agrochemicals Detection, Treatment and Remediation, Pesticides and Chemical Fertilizers* (pp. 621–645). Elsevier BV. 10.1016/b978-0-08-103017-2.00023-4.

Ashiq, A., Adassooriya, N. M., Sarkar, B., Rajapaksha, A. U., Ok, Y. S., & Vithanage, M., (2019). Municipal solid waste biochar-bentonite composite for the removal of antibiotic ciprofloxacin from aqueous media. *J. Environ. Manag., 236*, 428–435. https://doi.org/10.1016/j.jenvman.2019.02.006.

Awasthi, G., Nagar, V., Mandzhieva, S., Minkina, T., Sankhla, M. S., Pandit, P. P., Aseri, V., Awasthi, K. K., Rajput, V. D., Bauer, T., et al., (2022). Sustainable amelioration of heavy metals in soil ecosystem: Existing developments to emerging trends. *Minerals, 12*, 85. https://doi.org/10.3390/min12010085.

Cao, Y., Zhang, S., Zhong, Q., Wang, G., Xu, X., Li, T., Wang, L., et al., (2018). Feasibility of nanoscale zero-valent iron to enhance the removal efficiencies of heavy metals from polluted soils by organic acids. *Ecotoxicol. Environ. Saf., 162*, 464–473. 10.1016/j.ecoenv.2018.07.036.

Chaudhary, P., Khati, P., Chaudhary, A., Maithani, D., Kumar, G., & Sharma, A., (2021). Cultivable and metagenomic approach to study the combined impact of nanogypsum and *Pseudomonas taiwanensis* on maize plant health and its rhizospheric microbiome. *PLoS One, 16*, e0250574. https://doi.org/10.1371/journal.pone.0250574.

Chauhan, G., González-González, R. B., & Iqbal, H. M., (2021). Bioremediation and decontamination potentials of metallic nanoparticles loaded nanohybrid matrices—A review. *Environ. Res., 204*, 112407. 10.1016/j.envres.2021.

Chen, L., Chen, X. L., Zhou, C. H., et al., (2017). Environmental-friendly montmorillonite-biochar composites: Facile production and tunable adsorption-release of ammonium and phosphate. *J. Clean. Prod., 156,* 648–659. https://doi.org/10.1016/j.jclepro.2017.04.050.

Chen, W. H., Hsu, H. C., Lu, K. M., Lee, W. J., & Lin, T. C., (2011). Thermal pretreatment of wood (Lauan) block by torrefaction and its influence on the properties of the biomass. *Energy, 36*, 3012–3021. 10.1016/j.energy.2011.02.045.

Dagar, J. C., Yadav, R. K., & Sharma, P. C., (2019). *Research Developments in Saline Agriculture.* Springer. https://doi.org/10.1007/978-981-13-5832-6.

Davies, B., Coulter, J. A., & Pagliari, P. H., (2022). Soil enzyme activity behavior after urea nitrogen application. P*lants, 11*, 2247. https://doi.org/10.3390/ plants11172247.

El-Saadony, M. T., Saad, A. M., Soliman, S. M., Salem, H. M., Desoky, E. S. M., Babalghith, A. O., El-Tahan, A. M., et al., (2022). Role of nanoparticles in enhancing crop tolerance to abiotic stress: A comprehensive review. *Front. Plant Sci., 13*, 946717. doi: 10.3389/fpls.2022.946717.

Ghollarata, M., & Raiesi, F., (2007). The adverse effects of soil salinization on the growth of *Trifolium alexandrinum* L. and associated microbial and biochemical properties in soil from Iran. *Soil Biology and Biochemistry, 39*, 1699–1702. https://doi.org/10.1016/j.soilbio.2007.01.024.

Greco, C., Cosentino, U., Pitea, D., Moro, G., Santangelo, S., Patan`e, S., D'Arienzo, M., et al., (2019). Role of the carbon defects in the catalytic oxygen reduction by graphite nanoparticles: A spectromagnetic, electrochemical and modelistic integrated approach. *Phys. Chem. Chem. Phys., 21*, 6021–6032.

Hosny, M., Fawzy, M., & Eltaweil, A. S., (2022). Green synthesis of bimetallic Ag/ZnO@Biohar nanocomposite for photocatalytic degradation of tetracycline, antibacterial and antioxidant activities. *Scientific Reports, 12*, 7316. https://doi.org/10.1038/s41598-022-11014-0.

Jamil, A., Riaz, S., Ashraf, M., & Foolad, M. R., (2011). Gene expression profiling of plants under salt stress. *Crit. Rev. Plant Sci., 30*(5), 435–458. https://doi.org/10.1080/07352689.2011.605739.

Jiang, M., He, L., Niazi, N. K., Wang, H., Gustave, W., Vithanage, M., Geng, K., et al., (2023). Nanobiochar for the remediation of contaminated soil and water: Challenges and opportunities. *Biochar, 5*, 2. https://doi.org/10.1007/s42773-022-00201-x.

Khader, A., Ibrahim, M., Alkhathami, F., Elsawy, H., El-Kader, N. A., Shaker, E., & Mahmoud, E., (2023). Properties of nano-amendments and their effect on some soil properties and root-knot nematode and yield attributes of tomato plant. *Agriculture, 13*(2), 366.

Khaled, H., & Fawy, H. A., (2011). Effect of different levels of humic acids on the nutrient content, plant growth, and soil properties under conditions of salinity. *Soil & Water Res., 6*(1), 21–29. https://doi.org/10.17221/4/2010-SWR.

Khan, H. A., Naqvi, S. R., Mehran, M. T., Khoja, A. H., Niazi, M. B. K., Juchelková, D., & Atabani, A., (2021). A performance evaluation study of nano-biochar as a potential slow-release nano-fertilizer from wheat straw residue for sustainable agriculture. *Chemosphere, 285*, 131382.

Khan, I., Seleiman, M. F., Chattha, M. U., Jalal, R. S., Mahmood, F., Hassan, F. A., et al., (2021). Enhancing antioxidant defense system of mung bean with a salicylic acid exogenous application to mitigate cadmium toxicity. *Not. Bot. Horti. Agrobot. Cluj Napoca., 49*, 12303.

Khan, S. T., Adil, S. F., Shaik, M. R., Alkhathlan, H. Z., Khan, M., & Khan, M., (2022). Engineered nanomaterials in soil: Their impact on soil microbiome and plant health. *Plants, 11*, 109. 10.3390/plants11010109.

Khanna, K., Kohli, S. K., Handa, N., Kaur, H., Ohri, P., Bhardwaj, R., Yousaf, B., et al., (2021). Enthralling the impact of engineered nanoparticles on soil microbiome: A concentric approach towards environmental risks and cogitation. *Ecotoxicol. Environ. Saf., 222*, 112459. https://doi.org/10.1016/j.ecoenv.2021.112459.

Kumari, S., Khanna, R. R., Nazir, F., Albaqami, M., Chhillar, H., Wahid, I., et al., (2022). Bio-synthesized nanoparticles in developing plant abiotic stress resilience: A new boon for sustainable approach. *Int. J. Mol. Sci., 23*, 4452. doi: 10.3390/ijms23084452.

Leogrande, R., & Vitti, C., (2019). Use of organic amendments to reclaim saline and sodic soils: A907review. *Arid L. Res. Manag., 33*(1), 1–21. https://doi.org/10.1080/15324982.2018.1498038.

Lobzenko, I., Burachevskaya, M., Zamulina, I., Barakhov, A., Bauer, T., Mandzhieva, S., Sushkova, S., Minkina, T., Tereschenko, A., Kalinichenko, V., et al., (2022). Development of a unique technology for the pyrolysis of rice husk biochar for promising heavy metal remediation. *Agriculture, 12*, 1689. https://doi.org/10.3390/agriculture12101689.

Naghdi, M., Taheran, M., Pulicharla, R., et al., (2019). Pine-wood derived nanobiochar for removal of carbamazepine from aqueous media: Adsorption behavior and influential parameters. *Arab J. Chem., 12*, 5292–5301. https://doi.org/10.1016/j.arabjc.2016.12.025.

Neeli, S. T., & Ramsurn, H., (2018). Synthesis and formation mechanism of iron nanoparticles in graphitized carbon matrices using biochar from biomass model compounds as a support. *Carbon, 134*, 480–490. https://doi.org/10.1016/j.carbon.2018.03.079.

Noreen, S., & Abd-Elsalam, K. A., (2021). Biochar-based nanocomposites: A sustainable tool in wastewater bioremediation. In: Abd-Elsalam, K. A., & Zahid, M., (eds.), *Applications of Nanomaterials for Water Purification, Micro and Nano Technologies, Aquananotechnology* (pp. 185–200). Elsevier BV.

Parkash, V., & Singh, S., (2020). Potential of biochar application to mitigate salinity stress in eggplant. *HortScience, 55*(12), 1946–1955.

Patel, A., Khare, P., & Patra, D. D., (2017). Biochar mitigates salinity stress in plants. In: *Plant Adaptation Strategies in Changing Environment.* https://doi.org/10.1007/978-981-10-6744- 0_6.

Pratap, T., Chaubey, A. K., Patel, M., Mlsna, T. E., Pittman, C. U. Jr., & Mohan, D., (2022). Nanobiochar for aqueous contaminant removal. In: Mohan, D., Pittman, C. U., & Mlsna, T. E., (eds.), *Sustainable Biochar for Water and Wastewater Treatment* (pp. 667–704). Elsevier, Amsterdam. https://doi.org/10.1016/B978-0-12-822225-6.00021-X.

Prăvălie, R., Patriche, C., Borrelli, P., Panagos, P., Roşca, B., Dumitraşcu, M., Nita, I. A., et al., (2021). Arable lands under the pressure of multiple land degradation processes. A global perspective. *Environmental Research, 194*, 110697. https://doi.org/10.1016/j.envres.2020.110697.

Qian, Y., Qin, C., Chen, M., & Lin, S., (2020). Nanotechnology in soil remediation—Applications vs. implications. *Ecotoxicol. Environ. Saf., 201*, 110815. 10.1016/j.ecoenv.2020.110815.

Quilliam, R. S., Glanville, H. C., Wade, S. C., & Jones, D. L., (2013). Life in the 'charosphere' – does biochar in agricultural soil provide a significant habitat for microorganisms? *Soil Biol. Biochem., 65*, 287–293. https://doi.org/10.1016/j.soilbio.2013.06.004.

Rabbani, M. M., Ahmed, I., & Park, S. J., (2016). Application of nanotechnology to remediate contaminated soils. In: *Environmental Remediation Technologies for Metal-Contaminated Soils* (pp. 219–229). Springer: Tokyo, Japan. 10.1007/978-4-431-55759-3_10.

Rajput, V. D., Chen, Y., & Ayup, M., (2015). Effects of high salinity on physiological and anatomical indices in the early stages of *Populus euphratica* growth. *Russ. J. Plant Physiol., 62*, 229–236. https://doi.org/10.1134/S1021443715020168.

Rajput, V. D., Kumari, A., Upadhyay, S. K., Minkina, T., Mandzhieva, S., Ranjan, A., Sushkova, S., Burachevskaya, M., Rajput, P., Konstantinova, E., et al., (2023). Can nanomaterials improve the soil microbiome and crop productivity? *Agriculture, 13*, 231. https://doi.org/10.3390/ agriculture13020231.

Ramanayaka, S., Vithanage, M., Alessi, D., Liu, W. J., Jayasundera, A. C. A., & Ok, Y. S., (2020). Nanobiochar: Production, properties, and multifunctional applications, *Environ. Sci. Nano, 7*, 3279–3302.

Rani, S., Sharma, M. K., Kumar, N., & Neelam, (2019). Impact of salinity and zinc application on growth, physiological and yield traits in wheat. *Current Science, 116*(8), 1324–1330. 10.18520/cs/v116/i8/1324-1330.

Rashid, M. I., Shah, G. A., Sadiq, M., Amin, N. U., Ali, A. M., Ondrasek, G., & Shahzad, K., (2023). Nanobiochar and copper oxide nanoparticles mixture synergistically increases soil nutrient availability and improves wheat production. *Plants, 12*(6), 1312.

Rengasamy, P., (2006). World salinization with emphasis on Australia. In: *Journal of Experimental Botany*. https://doi.org/10.1093/jxb/erj108.

Sahab, S., Suhani, I., Srivastava, V., Chauhan, P. S., Singh, R. P., & Prasad, V., (2021). Potential risk assessment of soil salinity to agroecosystem sustainability: Current status and management strategies. *Science of The Total Environment, 764,* 144164. https://doi.org/10.1016/j.scitotenv.2020.144164.

Saxena, J., Rawat, J., & Kumar, R., (2017). Conversion of biomass waste into biochar and the effect on mung bean crop production. *Clean, 45*, 1501020. 10.1002/clen.201501020.

Seleiman, M. F., Semida, W. M., Rady, M. M., Mohamed, G. F., Hemida, K. A., Alhammad, B. A., et al., (2020). Sequential application of antioxidants rectifies ion imbalance and strengthens antioxidant systems in salt-stressed cucumber. *Plants, 9*, 1783. 10.3390/plants9121783.

Sharma, N., Singh, G., Sharma, M., Mandzhieva, S., Minkina, T., & Rajput, V. D., (2022). Sustainable use of nano-assisted remediation for mitigation of heavy metals and mine spills. *Water, 14*, 3972. https://doi.org/10.3390/w14233972.

Taha, R. S., Seleiman, M. F., Shami, A., Alhammad, B. A., & Mahdi, A. H. A., (2021). Integrated application of selenium and silicon enhances growth and anatomical structure, antioxidant defense system and yield of wheat grown in salt-stressed soil. *Plants, 10*, 1040. doi: 10.3390/plants10061040.

Tiwari, R., Dwivedi, B. S., Sharma, Y. M., Sharma, A., & Dwivedi, A. K., (2019). Activities of β-glucosidase, phosphatase and dehydrogenase as soil quality indicators: A review. *Int. J. Curr. Microbiol. App. Sci., 8*(6), 834–846. https://doi.org/10.20546/ijcmas.2019.806.101.

Tóth, G., Adhikari, K., Várallyay, G., Tóth, T., Bódis, K., & Stolbovoy, V., (2008). Updated map of salt affected soils in the European union. *Threats to Soil Quality in Europe* (65–77). EUR 23438 EN.

Verma, K. K., Song, X. P., Joshi, A., Tian, D. D., Rajput, V. D., Singh, M., Arora, J., et al., (2022). Recent trends in nano-fertilizers for sustainable agriculture under climate change for global food security. *Nanomaterials, 12*, 173. https://doi.org/10.3390/nano12010173.

Xiao, Y., Lyu, H., Tang, J., Wang, K., & Sun, H., (2020). Effects of ball milling on the photochemistry of biochar: Enrofoxacin degradation and possible mechanisms. *Chem. Eng. J., 384*, 123311. 10.1016/j.cej.2019.123311.

Yadav, S. P., Bharadwaj, R., Nayak, H., Mahto, R., Singh, R. K., & Prasad, S. K., (2019). Impact of salt stress on growth, productivity and physicochemical properties of plants: A review. *Int. J. Chem. Stud., 7,* 1793–1798.

Zaman, M., Shahid, S. A., & Heng, L., (2018). *Guideline for Salinity Assessment, Mitigation and Adaptation Using Nuclear and Related Techniques.* https://doi.org/10.1007/978-3-319-96190-11253.

Zhang, X., Wells, M., Niazi, N., Bolan, N., Shaheeng, S., Hou, D., Gao, B., et al., (2022). Nanobiochar-rhizosphere interactions: Implications for the remediation of heavy-metal contaminated soils. *Environ. Pollut., 299,* 118810. 10.1016/j.envpol.2022.118810.

Zhao, F., Xin, X., Cao, Y., Su, D., Ji, P., Zhu, Z., & He, Z., (2021). Use of carbon nanoparticles to improve soil fertility, crop growth and nutrient uptake by corn (*Zea mays* L.). *Nanomaterials, 11,* 2717. 10.3390/nano11102717.

CHAPTER 8

Role of Nano-Biochar for Improving Soil Health

SOMAYEH RAHMAT

Department of Botany, Payame Noor University, Miandoab, Iran

ABSTRACT

Soil is a limited and fragile resource for agricultural products. Soil health is defined as "the continued capacity of the soil to function as a vital living ecosystem that sustains plants, animals, and humans." Therefore, maintaining it is very important. Nowadays, with the increasing global demand for production, the importance of soil health is doubled. However, various human activities seriously damage it. This, coupled with a lack of proper and sustainable management, leads to soil erosion, reduction of soil fertility, decreased plant and crop growth, and damage to food security. In such cases, one effective strategy is to add amendments made from biomass residue called Nano Biochar (N-BC), which persists in the soil for hundreds and even thousands of years and improves the physical, hydraulic, chemical, and biological properties of the soil. Biochar has long been known as an effective soil conditioner. Reducing the particle size of biochar to the nanoscale, due to its higher surface-to-volume ratio, surface energy, surface adsorption potential, and subsequently its biological effectiveness compared to fresh biochar, leads to a more efficient soil amendment. Depending on the soil structure and biochar variables, N-BC can increase soil cation exchange capacity (CEC), pH, organic matter, water holding capacity (WHC), nutrient profile, soil structure, microbial activity, and gas exchange properties. Additionally, N-BC can reduce the mobility/bioavailability of various types of soil pollutants. As a result, nano-BC suggests a potentially sustainable,

Nanomaterials and Nano-Biochar in Reducing Soil Stress: An Integrated Approach to Sustainable Agriculture. Vishnu D. Rajput, & Tatiana Minkina (Eds.)

affordable, and promising technology for the proper management of soil and agricultural waste. It also ensures food security and the development of sustainable and environmentally friendly agriculture. However, it is highly recommended to conduct scientific and industrial additional studies to investigate the potential environmental risks caused by the small size of N-BC before further development.

8.1 INTRODUCTION

Soil can be considered as one of the constrained and fragile resources of agricultural production. Accordingly, one of the significant necessities for sustainable food production and security is to maintain soil health and productivity. The concept of soil health has developed since the late 1990s (Hou et al., 2020; Doran, 1996) in order to guarantee the economical management and use of this valuable asset. Soil health is defined as "the soil's capacity to act as a biological system that supports plants, humans, and animals." A healthy soil maintains biodiversity, which plays a critical role in controlling plant diseases and pests, recycling essential nutrients for plant growth, forming useful symbiotic bonds with plant roots, improving soil structure to conserve water and nutrients, and ultimately enhancing plant yield (FAO, 2011). "Soil quality" and "Soil health" are generally considered synonymous. Hence, they can be utilized interchangeably, with one key refinement being conceptualized over the past decades: soil quality accounts for the inborn and energetic quality; intrinsic soil quality speaks to those aspects of the soil quality that are associated with the soil composition and natural characteristics [soil type (Natural Resources Conservation Services: Soil Health, 2012)], being affected by long-term natural agents and the process of soil formation. These characteristics are generally not influenced by human intervention. On the other hand, dynamic soil quality, which is considered equivalent to soil health, reflects those characteristics of the soil that change as a result of soil use and management on a human time scale (Moebius-Clune, 2016).

In advanced society, various human activities such as erosion (Abd El-Halim & Omae, 2019; Boardman et al., 2019; Tao et al., 2019), acidification (Liao et al., 2019; Patriche, 2019), contamination (Hou, 2020; Wang et al., 2021), and compaction (Daraghmeh et al., 2019; Tassinari et al., 2019) have damaged soil health. Typically, it is expected that more food will be produced sustainably on healthy soils based on soil preservation and reclamation techniques in order to feed 9.7 billion people by the year 2050 (Godfray et al., 2010). One of the cost-effective and environmentally friendly amendment approaches is the application of Biochar as a conditioner with a long history, which ensures sustainable crop production in numerous ways. Biochar is a carbon-rich black solid that is produced from a wide range of natural wastes such as biosolids, agricultural residues, manure, green wastes, and animal carcasses (Bolan et al., 2022) through various controlled thermochemical processes, such as quick and slow pyrolysis, flash carbonization, hydrothermal carbonization, gasification, and torrefaction (Lehmann, 2007; Meyer et al., 2011). Renewable energy resulting from biomass combustion, such as the combustion of plant residues, is ultimately carbon neutral, and the emitted CO_2 is ultimately consumed by plants. In the case of biochar, despite its organic and natural origin, due to the high proportion of dense aromatic carbon, high porosity, and large surface area (SA), it is extremely stable and can be deposited for centuries or even millennia (Chausali et al., 2021). This period has been estimated to be up to 2000 years for biochar prepared from ^{14}C labeled perennial ryegrass (Kuzyakov et al., 2009). These properties are associated with the agricultural and environmental benefits of applying biochar to the soil. In fact, it is suggested to consider biochar as a soil amendment to reduce climate change through carbon sequestration, as well as to decrease greenhouse gas emissions (Kammann et al., 2012; Mukherjee & Lal, 2013; Woolf et al., 2010). It is also a very good option for enhancing soil health and promoting plant growth through several mechanisms associated with improving soil structure, porosity, pH, organic carbon concentration, water and nutrient retention, CEC, and WHC (Biederman & Harpole, 2013; Demiraj et al., 2018; Domingues et al., 2020; Githinji, 2014; Libutti et al., 2016; Herath et al., 2013; Libutti et al., 2019, 2021). Additionally, biochar can enhance the activity and growth of beneficial microbial communities, as well as soil enzyme activity and suppress soil pathogens (Altland & Locke, 2017; Bonanomi et al., 2017; Hussain et al., 2017; Jones et al., 2012; Murtaza et al., 2021; Rehman et al., 2021). Moreover, biochar reduces the mobility/bioavailability of various types of pollutants, such as heavy metals (HMs), in contaminated soils, resulting in an effective action to restore the efficiency of degraded agricultural soils (Ćwieląg-Piasecka et al., 2018; Graber et al., 2010; Sajjadi et al., 2019; Zama et al., 2018). Several parameters, such as the origin of biochar (Lu et al., 2017), pyrolysis temperature (Yue et al., 2019), heating time (Bruun et al., 2012), application rate (Cayuela et al., 2013; Frenkel et al., 2017; Nguyen et al., 2017), soil type and texture (Blanco-Canqui, 2017), and time (Cayuela et al., 2013) and form of application (Borchard et al., 2014; Karami et al., 2011; Sadegh-Zadeh et al., 2018), control the efficiency of biochar in various agricultural applications.

Nanotechnology has been suggested as an alluring approach to improve surface characteristics, including specific surface area (SSA), surface porosity, and surface functional groups. Various methods are used to convert bulk biochar (B-BC) into N-BC with more favorable characteristics as well as more extensive applications in agricultural, environmental, and other areas (Pascoli et al., 2019; Wang & Nguyen, 2018). N-BC is tested or used for soil amendment for several reasons: First, the formation of N-BC is unavoidable during the production, relocation, and implementation of B-BC (Song et al., 2022); for example, N-BC can account for 0.87% to 2.36% of the whole biochar during peanut shell pyrolysis, probably via matrix fracture and pore collapse pathways (Liu et al., 2018). After spreading in the soil and air, these particles enter the waters, then spread to the entire ecosystem (Song et al., 2022). Second, since the structure, composition, and properties of N-BC are diverse from B-BC, its effect on plant growth and its function in soil amendment are probably different from the bulk type (Song et al., 2022). Third, most parts of biochar can even be decomposed into N-BC because of aging and weathering in the environment (Spokas et al., 2014). However, few studies have addressed the effect of N-BC on soil characteristics. This chapter tries to summarize the existing literature on the effect of N-BC on soil health, including physicochemical and biological properties. Environmental concerns of N-BC are also discussed.

8.2 WHAT IS N-BC?

Both B-BC and N-BC can be obtained from biochar produced by carbonizing feedstock under anoxic or oxygen-limited conditions. The main difference between B-BC and N-BC is the difference in particle size (Song et al., 2022). B-BCs have dimensions ranging from 0.04 to 20 millimeters (Gao & Wu, 2014). During N-BC production, micronized biochar includes both parts with dimensions less than one micrometer (μm) to nanometer (nm), so that dimensions up to 100 nm are defined as colloidal, and those less than 100 nm as N-BC (Liu et al., 2018). In two recent studies, the preparation of N-BC with dimensions less than 5 nm has been reported (Lian et al., 2018; Xiao & Chen, 2017). There are various techniques for the production of N-BC in the literature, including ball milling (Law et al., 2020; Ramezanzadeh et al., 2021), carbonization (Makshut et al., 2020; Zhang et al., 2021), ultrasound

(Liu et al., 2018; Oleszczuk et al., 2016), centrifugation (Yao et al., 2020; Song et al., 2019), and grinding (Chen et al., 2020). Ball milling is the method that is most interesting to researchers for producing N-BC (Kumar et al., 2020). The characteristics of N-BC are a direct function of the type of feedstock and the conditions applied to the parent biochar (Kim et al., 2020). Based on FTIR results, N-BC prepared from lignin is different from the bulk type by higher density and various types of surface functional groups such as carbonyl, carboxyl, and hydroxyl, aromaticity, less C-C bonds, and more polarity (Ramezanzadeh et al., 2021; Song et al., 2019; Jiang et al., 2020). Besides, for the same raw materials, increasing the pyrolysis temperature can decrease the size of the obtained N-BC particles. For example, expanding the pyrolysis temperature from 400 to 700°C for rice straw diminished the molecule estimate from 403 nm to 234 nm by using the centrifuge separation method (Lian et al., 2020). Yue et al. also confirmed the result as well as reported that the N-BC particles obtained at higher temperatures had smaller size, more developed pore structure, and higher Si content than N-BC particles produced at lower temperatures, which indicates their higher surface reactivity (Yue et al., 2019).

A diminish within the measure of biochar particles to 100 nm or less can be considered as an effective approach for increasing the surface energy, adsorption potential, and thus biological effectiveness with a higher surface-area-to-volume ratio, and also promotes suboptimal biochar into high-quality nanomaterials called the N-BC (Weber & Quicker, 2018). It has been shown that size reduction can cause changes in various parameters that are responsible for the distinct characteristics of N-BC in comparison to B-BC (Song et al., 2022). N-BC is characterized via significant physical and chemical differences, e.g., chemical composition, SSA, CEC, pH, pore characteristics, different zeta potential, and carbon defects that can produce ROS and as a result, increase the absorption capacity compared to the non-nano counterparts and also has characteristics more similar to carbon-based materials such as black carbon (Elsawy et al., 2022; Lyu et al., 2018; Oleszczuk et al., 2016; Qin et al., 2018;). N-BC X-ray diffraction analysis revealed the existence of Mg-calcite mineral with turbostratic crystals (Ramezanzadeh et al., 2021). The comparison of the elemental composition of B-BC and N-BC derived from it also suggested the presence of phenolic components in the N-BC structure (Zhang et al., 2019). Therefore, accurate identification of N-BC properties plays an important role in the production of N-BCs with specific practical purposes.

8.3 EFFECTS OF N-BC ON SOIL HEALTH

Today's society is facing soil degradation and crop yield reduction due to excessive use of soil and climate change. The biochar potential in the enhancement of soil health and fertility has been widely and verifiably documented (Chen et al., 2019; Biswas et al., 2019; Wu et al., 2019), especially when added to soils with low fertility and acidic soils such as humid tropical areas (Berek et al., 2018; Jeffery et al., 2011). Despite the inherent properties of N-BC such as high surface area, microporosity, and stability making it a more attractive additive than biochar for improving soil quality and biological processes, there are few reports in this field and knowledge about the effects of N-BC on soil health is insignificant. The soil response to biochar involves complex physical, chemical, and biological interactions, which, as mentioned, are different in response to a wide range of factors associated with nanoparticles and soil (Chen et al., 2020).

8.3.1 EFFECTS OF N-BC ON SOIL PHYSICOCHEMICAL CHARACTERISTICS

Utilization of N-BC improves soil physicochemical characteristics (Table 8.1). Considering the high resistance of biochar against microbial decomposition and its long stability in the soil, the first advantage of using N-BC is the long-term increase in the level of soil organic matter (SOM), resulting in the sustainable improvement of soil properties. Adding biochar to soil due to the difference in physical structure between soil and biochar leads to changes in bulk density, tensile strength, hydrodynamics, and gas flow of soil (Abed et al., 2022). Also, due to the larger surface area (SA), higher surface charge density, more diverse crystalline forms, and more negative zeta potential than biochar, N-BC can easily bond with cations, anions, metals, and soil components on its surface. It can therefore act well as a slow-release fertilizer (Guo et al., 2013; Yang et al., 2020), improve the soil's ability to retain water and nutrients, and reduce leaching of nutrients and fertilizers (Chen et al., 2020; Elsawy et al., 2022; Kumar et al., 2020; Yang et al., 2020). On the other hand, these characteristics of N-BC cause the adsorption of heavy metals (HMs) and soil pollutants, thus improving soil fertility and health (Yue et al., 2019; Ramezanzadeh et al., 2021) (which will be discussed separately in the following). Moreover, N-BC can provide suitable physical conditions for plants and soil microorganisms by improving soil structure

TABLE 8.1 The Effect of N-BC on Physical, Chemical and Biological Characteristics of Soil

Feedstock Synthesis method	Particle size (nm)	BC temp	Focus of study.	Performance	References
Wheat/straw/-Pyrolysis	–	350–550°C	Effect on Phytotoxicity, soil remediation and microbial community in Cd (II) -contaminated soil	Reducing the Cd (II) content of the soil, Improving the diversity, abundance and biomass of actinobacteria and Bacteroidetes, Reducing the diversity of proteobacteria	Liu et al. (2020)
Rice-hull/ centrifuged at 10,000 rpm	-	300–600°C	Effect on phytotoxicity of Cd (II) in hydroponic condition	Significant reduction in Cd (II) bioavailability	Yue et al. (2019)
Coconut shell/ Ultrafine grinding	40–200	-	Effect on soil erosion and nitrate loss	Reduced sediment yield (8.9–41.8%), Reduced runoff (39.7–74.4%), Reduced nitrate loss (13.6–59.8%)	Chen et al. (2020)
Rice straw Rice straw Wheat straw Wheat straw Corn straw Corn straw/ Planetary Ball Mill	253.2 201.7 284.6 224.2 259.7 252.1	350°C 650°C 350°C 650°C 350°C 650°C	Protecting native species from invasive species	Stabilizing ferulic acid and reducing its allelopathic effects on rice plants	Shen et al. (2020)
Wood/ Ball milling	<412.4	100–650°C	Effect on Cd (II) sorption in sandy soil	Treating soils by N-BC have better performance than no or macrobiochar in cleaning environment polluted by Cd (II)	Ramezanzadeh et al. (2021)
Coconut shells/ Ultrafine grinding	40	-	Effect on soil water and physicochemical characteristics, fertilizer utilization efficiency	Reducing K, P and N leaching, Improving soil water and nutrient retention, Increasing aggregates with less than 2 mm particle size	Yang et al. (2020)
Wood sawdust/ Crushing with a mill (DING CANG DC-500A)	70–500	350°C	Effect on phosphorus absorption in soils under salinity stress	Increase the P absorption capacity up to 20% through surface precipitation mechanism	Mahmoud et al. (2020)
Oak tree branches/ – Pyrolysis	<40	500°C	Effect on losses of soil, nutrients and water	Reducing soil erosion, reducing water and nutrient losses. The significant effect of vegetation cover and N-BC content on soil water and nutrients retention	Patriche (2019)

and aggregation (Elsawy et al., 2022; Lian et al., 2018; Liu et al., 2020). Lian et al. stated that N-BC has a negative charge and it can be predicted that if N-BC with sufficient negative charge attaches to soil minerals and reverses their surface charge, it can increase mineral dispersion (Lian et al., 2018). In this regard, the effects of N-BC synthesized from peanut shells on the dispersion of natural soil inorganic nanoparticles (hematite, goethite, kaolinite, and montmorillonite) from soil to the aqueous phase were investigated. It was found that N-BC did not affect the mobility of montmorillonite and kaolinite. On the contrary, B-BC was able to reduce the dispersion of these aluminosilicates in the aqueous medium (Liu et al., 2018). Further investigations are needed to determine the effect of N-BC on other soil compounds such as organic matter and soluble salts. Other research on the effect of N-BC on soil properties indicates that, similar to the bulk type, N-BC increases the capillary force and WHC (Abed et al., 2022; Guo et al., 2013; Song et al., 2019). Due to its higher SSA, a lower rate of N-BC is required for soil water retention compared to B-BC (Tan et al., 2016). In Chen's et al. study, N-BC derived from coconut shells (1%) effectively decreased runoff, sedimentation, and leaching of nitrate. Interestingly, nitrate losses were negatively correlated with increased N-BC content (Chen et al., 2020). Additionally, low application rates of N-BC (0.5–1%) resulted in nitrate fixation comparable to higher application rates of biochar (0.8–2%) (Feng et al., 2018; Hailegnaw et al., 2019; Khare, 2021). The use of N-BC to control other macronutrients has also been reported. Zhou et al. investigated the impacts of adding N-BC (0–1%) to the soil during the development of seedlings of several plants. They found that N-BC limited the loss of nutrients (K, N, P) due to soil erosion for all the plants under manufactured rainfall conditions (Zhou et al., 2020). The application of biochar derived from pepperwood through an adsorption mechanism resulted in a decrease in nitrate, phosphate, and ammonium in the leachate fraction by 34%, 20.6%, and 34.7% respectively (Mukherjee et al., 2020). In one study, the effect of exposure to N-BC (0–1%) within the soil on the yield and growth of maize was investigated. The included N-BC increased the content of nutrients (K, P, and N) in the leaves, stems, and roots of the maize plants, and the yield of the plant in the presence of N-BC (0.7% and 1%) increased by 4.4% and 5.7% respectively (Yang et al., 2020).

Moreover, Mahmoud et al. investigated the effect of N-BC on P adsorption capacity in the soil under different levels of salinity stress. They identified that in the presence of N-BC, the bioavailable soil P increased, and there was a positive correlation between the P adsorption capacity value and the increased N-BC rate. The mechanism responsible for the adsorption of

P on N-BC, surface precipitation, was identified (Mahmoud et al., 2020). Additionally, due to the colloidal stability and high affinity of N-BC, the absorption and accumulation of trace elements in the soil can significantly increase (Song et al., 2019). Hossein and Siddiqi reported that the application of nanoparticles as nano-insecticides and fertilizers in degraded soils can enhance yield, decrease the excessive use of chemical fertilizers, and increase the vital capacity of the soil (Husen & Siddiqi, 2014). N-BC fertilizers can increase crop yield by 10% to 20% while decreasing the required fertilizer consumption by 30% to 50%, respectively (Wang et al., 2016). Considering that a significant portion of nitrogen fertilizers is converted into N_2O by nitrification and denitrification processes (Harter et al., 2014), it can be said that not only by significantly reducing the amount of CO_2 emissions through increasing the stability of soil organic carbon and reducing its mineralization (Ameloot et al., 2013), but also by reducing the input of nitrogen fertilizer to agricultural ecosystems and improving the efficiency of its utilization, N-BC can also be considered as a great apparatus to reduce N_2O outflow and consequently mitigate global warming (Harter et al., 2014). The processes of nitrification and nitrogen mineralization play a critical role in nitrate leaching. It has been reported that charcoal soil application can inhibit nitrate leaching by minimizing it (Wang et al., 2018). Despite these promising results, the side effects of N-BC soil application will be discussed in the following sections.

8.3.2 EFFECTS OF N-BC ON REMOVAL SOIL POLLUTANTS

Advanced structural nanomaterials, such as N-BC, offer verifiable solutions to some of the modern world's problems (Rajput et al., 2022). Considering its fine unique characteristics, i.e., larger SSA, surface hydrophobicity, and microporosity compared to the conventional biochar-based methods, N-BC has an extraordinary capacity to adsorb and remediate a wide range of pollutants, including HMs, herbicides, polycyclic aromatic hydrocarbons (PAHs), and Polychlorinated Biphenyls (PCBs) in different environments and has the capability of mobility. Given their higher pH and alkaline nature, they can be utilized to deal with soil acidity (Gámiz et al., 2017; Wang et al., 2013). Few researches have addressed the effect of N-BC on the remediation of soil pollutants (Table 8.1). Heavy metals contamination in soil is persistent (Pérez-Esteban et al., 2014) and can affect soil fertility, biodiversity, microbial activities, crop yield, as well as create risks to human health through diet

(Bolan et al., 2014; Ma et al., 2015). The application of N-BC can decrease the mobility/bioavailability and consequently the adsorption of potentially toxic metals in contaminated soils (Liu et al., 2020). The increased pH of the soil after the addition of N-BC can play a significant role in this process (Kätterer et al., 2019). Yue et al. indicated that the N-BC outperforms (at least 10 times higher affinity) the B-BC in inhibiting Cd (II) uptake by rice. They also found that the N-BC significantly decreases the uptake and toxicity of Cd (II) for the rice plants in the hydroponic environment by stimulating the antioxidant system (Yue et al., 2019). In another study, the N-BC (100 mg L^{-1} or higher) improved the yield of *Brassica Chinese* by decreasing the exchangeable forms of Cd (II) and its bioavailability, and increased the biomass and microbial diversity in the soil (Liu et al., 2020), which are critical indicators of soil health (Rajput et al., 2022). In addition, the results of adding 2% of B-BC and N-BC to sandy soil samples showed the maximum adsorption of Cd (II) at the rate of 328.9 and 1062.4 mg kg^{-1}, which was 58.6% and 412.2% more than the control soil, respectively (Ramezanzadeh et al., 2021). The accumulation of allelochemicals in the soil can be considered as one of the other main bottlenecks for the optimal utilization of the soil in agricultural lands (Chen et al., 2022). It causes interference with the growth of adjacent or consecutive plants, disturbs the balance of the soil microbial community, and decreases the crop yield and the rate of seed germination as well as the soil fertility (Wang et al., 2013). The use of N-BC prepared from wheat (200–250 nm) showed that N-BC was able to adsorb and stabilize ferulic acid in the soil and reduce its allelopathic impact on rice plants, which was reported to be beneficial for plant growth (Shen et al., 2020). In addition, due to its colloidal stability, N-BC is capable of forming aggregates with inorganic colloids, increasing their dispersion and having a significantly high affinity to phenanthrene (Lian et al., 2018). Belonging to the group of PAHs, phenanthrene is one of the organic pollutants that its removal from the soil is important due to its dangers.

8.3.3 EFFECTS OF N-BC ON SOIL BIOLOGICAL CHARACTERISTICS

There are numerous components in soil that form an open biochemical system. Microbial community plays a significant part in checking the environment function and health, so that any change within the microbial community impacts the whole biological system (Meier et al., 2017). Soil microorganisms play a primary role in organic matter, decomposition, nitrogen fixation, nutrient cycle, and soil health. A large amount of research has proven that the application of biochar in the soil can increase the microbial biomass and affect the cycle of nutrients, the activity writing enzymes as well as the health of crops (Lehmann et al., 2011; Muhammad et al., 2014; Watzinger et al., 2014).

Agreeing with the soil health definition displayed, retaining and changing the soil microbiome can be considered as the key to improving soil health (Bolan et al., 2022), and the N-BC can accomplish this objective by adopting the following approaches. N-BC can influence the soil physicochemical characteristics, including the soil porosity, aggregate stability, tensile strength, and WHC. In this way, it also affects the soil biological community (Siedt et al., 2021). N-BC also affects the availability of soil macromolecules including N, P, and K, which can support the increase of microbial biomass. Furthermore, N-BC provides preventive microhabitats for various microorganisms, which can serve as a nitrogen and carbon source (Sarfraz et al., 2019), and in turn, will improve the soil activity. In addition, in one study, it has been determined that the source of labeled N in plants is biochar, indicating that N-BC can play a role in providing nutrients for the soil microbial community (Tan et al., 2017). It has been detailed that the application of N-BC has a positive effect on the immobilization of Cd (II) ions, making it unabsorbable for plants as well as a certain group of microbial community in soil under Cd (II) stress (Liu et al., 2020). N-BC decreased the diversity of *Proteobacteria* primarily existing within the contaminated locales compared to the control condition, while improving the number, biomass, and diversity of microbial communities belonging to *Bacteroidetes* and Actinobacteria, which are also beneficial for remediation of the contaminated soils and soil fertility, as measured by high-throughput sequencing (Liu et al., 2020).

These findings were supported by a previous study by Liao et al. (2019). These microorganisms can stick to the N-BC surface and gradually form biofilms and communities that, in addition to changing the structure of the community, attach the N-BC to the surface of the soil minerals. Therefore, the microbial composition can play a vital role in the N-BC absorption in the soil (Liu et al., 2020). Moreover, Elsawy et al. (2022) have recently explored the impacts of including N-BC on the yield of rapeseed and biological activity in degraded soils, demonstrating that the N-BC, in addition to improving the soil health, enhanced the biological activity (catalase and dehydrogenase) and microbial biomass carbon, probably because of the higher organic carbon, CEC and content of nutrients, which played a significant part in enhancing the yield of rapeseed. Enzyme activity can be considered as a reasonable index for biological activity, soil quality and fertility, and agricultural productivity (Liao et al., 2019). Dehydrogenases as intracellular enzymes are associated with soil microbiome. Therefore, dehydrogenase activity is one of the best tools in estimating stability of organic matter and soil condition in soils amended with biochar (Mierzwa-Hersztek et al., 2020).

8.4 N-BC ENVIRONMENTAL AND HEALTH CONCERNS

Although most studies have highlighted the advantages of N-BC, there are limited reviews that have identified its negative effects, including suppression of seed germination at certain N-BC concentrations (Liu et al., 2018, 2020), potential oxidative stress to plants (Yue et al., 2019), and possible loss of bound nutrients under simulated rainfall conditions (Zhou et al., 2020). In addition to the numerous benefits, higher activity is the major problem associated with N-BC, recently causing many environmental and health concerns (Ma et al., 2019). Considering the high mobility potential of N-BC as well as the higher pollutant absorption capacity compared to B-BC, biochar-contaminated nanoparticles, as carriers, can facilitate the transfer of pollutants, increasing the chances of water and soil contamination (Lian & Xing, 2017). As another reason for concern, a recent study has shown that the durability of a stable chemical with high ecological risks, trifluoropropylmethylsiloxane, can increase in soils with a total organic carbon content of $\geq$150 mg g^{-1} (Zhi et al., 2020). It is clear that the application of N-BC for the purpose of soil amendment will increase the spread and distribution of these hazardous chemicals along with their persistence. Following soil and groundwater contamination, N-BC eventually finds its way into the human food chain (Song et al., 2022). Although the health effects of N-BC have not been well defined, its potential toxic effects on the health of plants, humans, and animals cannot be ignored. Some recent studies indicate that chemical compounds of N-BC can raise concerns (Sigmund et al., 2017; Wang et al., 2017). With the current knowledge about the effects of fine particles of charcoal (Shao et al., 2016; Taner et al., 2013), black carbon (Kong et al., 2013; Sahu et al., 2014), and carbon nanotubes (Wan et al., 2017), exposure to N-BC can result in health effects for humans and animals, including the immune system, cardiovascular, respiratory, nervous, and skin problems (Long et al., 2020). However, N-BC has not attracted enough research attention for human and ecosystem health to determine its disadvantages and harms, making it necessary to widely examine the toxic effects of N-BC utilization in soil.

8.5 CONCLUSION AND FUTURE PERSPECTIVE

Having superior physicochemical characteristics compared to B-BCs, N-BCs have better efficiency in increasing soil stability, health, and productivity. These biological compounds have a fine structure and a high SSA. In general, N-BCs derived from different feedstocks and different production conditions have unique physicochemical properties with positive effects on all types of soils. However, at the same time, the wide variation in N-BC characteristics, despite its documented benefits on the soil, imposes limitations on its applications in agriculture. N-BC production is considered an effective and green technology that can improve various physical characteristics of the soil (e.g., aggregate stability, WHC, bulk density), chemical characteristics (e.g., CEC, nutrient profile, organic matter), and biological characteristics (e.g., microbial biomass, microbial richness, microbial diversity, enzyme activity), as well as its fertility. Additionally, N-BC can be used as a substrate for macro and micronutrients, allowing the nutrients to take the form of slow-release fertilizer by being placed in this substrate. The nutrients associated with N-BC can increase microbial diversity and plant yield by transferring to plants and microorganisms. Another advantage of N-BC is its capability in the remediation of various organic and inorganic pollutants from the soil. Despite its numerous points of interest, the N-BC's capability to move with surface runoff and groundwater makes it a potential carrier of pollutants. Therefore, the related risk to the health of soil life forms (i.e., microorganisms and soil fauna) and humans cannot be overlooked. The review of the existing literature on applications of N-BC in the soil revealed that there is a deep knowledge gap in this area. Many, if not all, of the N-BC compounds will remain in the soil for many generations after their addition. While most current studies on N-BC in soil have been limited to short-term laboratory incubation using relatively small sampling methods. Therefore, the investigation of the field and long-term effects, as well as the safety of N-BC applications in soil amendment on soil organisms, plants, and humans, is necessary. Another problem associated with the current studies is that although N-BC is characterized as a biochar with a particle measure of less than 100 nm, investigation writing has considered a wide range of particle sizes as N-BC. Such contradiction in particle estimate makes it troublesome to compare the yields of distinctive biochars, as the surface areas of the particles are generally impacted by their size. It is hoped that by providing comprehensive knowledge about the application of N-BC in soil and the use of such materials, fertile agricultural soils will be removed from the risk of destruction and erosion, and the production of agricultural products per unit area will increase.

KEYWORDS

- **microbial**
- **N-BC**
- **physicochemical**
- **soil characteristics**
- **soil health**

REFERENCES

Abd El-Halim, A. A., & Omae, H., (2019). Performance assessment of nanoparticulate lime to accelerate the downward movement of calcium in acid soil. *Soil Use and Management, 35*(4), 683–690.

Abed, H. B., Mahdi, A. B., Emad, I. S., Acwin, D. N. K., Romero, P. R. M., Barboza, A. L. A., & Kianfar, E., (2022). Production, structural properties nano biochar and effects nano biochar in soil: A review. *Egyptian Journal of Chemistry, 65*(12), 607–618.

Altland, J. E., & Locke, J. C., (2017). High rates of gasified rice hull biochar affect geranium and tomato growth in a soilless substrate. *Journal of Plant Nutrition, 40*(13), 1816–1828.

Ameloot, N., Graber, E. R., Verheijen, F. G., & De Neve, S., (2013). Interactions between biochar stability and soil organisms: Review and research needs. *European Journal of Soil Science, 64*(4), 379–390.

Beheshti, M., Etesami, H., & Alikhani, H. A., (2018). Effect of different biochars amendment on soil biological indicators in a calcareous soil. *Environmental Science and Pollution Research, 25*(15), 14752–14761.

Berek, A. K., Hue, N. V., Radovich, T. J. K., & Ahmad, A. A., (2018). Biochars improve nutrient phytoavailability of Hawaii's highly weathered soils. *Agron., 8*, 203–221.

Biederman, L. A., & Harpole, W. S., (2013). Biochar and its effects on plant productivity and nutrient cycling: A meta-analysis. *GCB Bioenergy, 5*(2), 202–214.

Biswas, B., Nirola, R., Biswas, J. K., Pereg, L., Willett, I. R., & Naidu, R., (2019). Environmental microbial health under changing climates: State, implication and initiatives for high-performance soils. In: *Sustainable Agriculture Reviews* (Vol. 29, pp. 1–32). Springer, Cham.

Blanco-Canqui, H., (2017). Biochar and soil physical properties. *Soil Science Society of America Journal, 81*(4), 687–711.

Boardman, J., Vandaele, K., Evans, R., & Foster, I. D., (2019). Off-site impacts of soil erosion and runoff: Why connectivity is more important than erosion rates. *Soil Use and Management*, *35*(2), 245–256.

Bolan, N., Hoang, S. A., Beiyuan, J., Gupta, S., Hou, D., Karakoti, A., & Van, Z. L., (2022). Multifunctional applications of biochar beyond carbon storage. *International Materials Reviews*, *67*(2), 150–200.

Bolan, N., Kunhikrishnan, A., Thangarajan, R., Kumpiene, J., Park, J., Makino, T., & Scheckel, K., (2014). Remediation of heavy metal (loid) s contaminated soils–to mobilize or to immobilize? *Journal of Hazardous Materials*, *266*, 141–166.

Bonanomi, G., Ippolito, F., Cesarano, G., Nanni, B., Lombardi, N., Rita, A., & Scala, F., (2017). Biochar as plant growth promoter: Better off alone or mixed with organic amendments? *Frontiers in Plant Science*, *8*, 1570.

Borchard, N., Siemens, J., Ladd, B., Möller, A., & Amelung, W., (2014). Application of biochars to sandy and silty soil failed to increase maize yield under common agricultural practice. *Soil and Tillage Research*, *144*, 184–194.

Bruun, E. W., Ambus, P., Egsgaard, H., & Hauggaard-Nielsen, H., (2012). Effects of slow and fast pyrolysis biochar on soil C and N turnover dynamics. *Soil Biology and Biochemistry*, *46*, 73–79.

Cayuela, M. L., Sánchez-Monedero, M. A., Roig, A., Hanley, K., Enders, A., & Lehmann, J., (2013). Biochar and denitrification in soils: When, how much and why does biochar reduce N_2O emissions? *Scientific Reports*, *3*(1), 1–7.

Chausali, N., Saxena, J., & Prasad, R., (2021). Nanobiochar and biochar based nanocomposites: Advances and applications. *Journal of Agriculture and Food Research*, *5*, 100191.

Chen, Q., Lan, P., Wu, M., Lu, M., Pan, B., & Xing, B., (2022). Biochar mitigates allelopathy through regulating allelochemical generation from plants and accumulation in soil. *Carbon Research*, *1*(1), 6.

Chen, W., Meng, J., Han, X., Lan, Y., & Zhang, W., (2019). Past, present, and future of biochar. *Biochar*, *1*(1), 75–87.

Chen, X., Zhou, B., Wang, Q., Tao, W., & Lin, H., (2020). Nano-biochar reduced soil erosion and nitrate loss in sloping fields on the Loess Plateau of China. *Catena*, *187*, 104346.

Ćwieląg-Piasecka, I., Medyńska-Juraszek, A., Jerzykiewicz, M., Dębicka, M., Bekier, J., Jamroz, E., & Kawałko, D., (2018). Humic acid and biochar as specific sorbents of pesticides. *Journal of Soils and Sediments*, *18*(8), 2692–2702.

Daraghmeh, O. A., Petersen, C. T., Munkholm, L. J., Znova, L., Obour, P. B., Nielsen, S. K., & Green, O., (2019). Impact of tillage intensity on clay loam soil structure. *Soil Use and Management*, *35*(3), 388–399.

Demiraj, E., Libutti, A., Malltezi, J., Rroço, E., Brahushi, F., Monteleone, M., & Sulçe, S., (2018). Effect of organic amendments on nitrate leaching mitigation in a sandy loam soil of Shkodra district, Albania. *Italian Journal of Agronomy*, *13*(1), 93–102.

Domingues, R. R., Sánchez-Monedero, M. A., Spokas, K. A., Melo, L. C., Trugilho, P. F., Valenciano, M. N., & Silva, C. A., (2020). Enhancing cation exchange capacity of weathered soils using biochar: Feedstock, pyrolysis conditions and addition rate. *Agronomy*, *10*(6), 824.

Doran, J. W., (1996). Soil health and sustainability. *Advances in Agron.*, *56*, 1–54.

Elsawy, H., El-shahawy, A., Ibrahim, M., El-Halim, A. E. H. A., Talha, N., Sedky, A., & Mahmoud, E., (2022). Properties of recycled nanomaterials and their effect on biological activity and yield of canola in degraded soils. *Agriculture*, *12*(12), 2096.

FAO, (2011). *Save and Grow – a Policymaker's Guide to the Sustainable Intensification of Smallholder Crop Production.* Rome: Food and Agriculture Organization of the United Nations.

Feng, Y., Sun, H., Xue, L., Wang, Y., Yang, L., Shi, W., & Xing, B., (2018). Sawdust biochar application to rice paddy field: Reduced nitrogen loss in floodwater accompanied with increased NH_3 volatilization. *Environmental Science and Pollution Research, 25*(9), 8388–8395.

Frenkel, O., Jaiswal, A. K., Elad, Y., Lew, B., Kammann, C., & Graber, E. R., (2017). The effect of biochar on plant diseases: What should we learn while designing biochar substrates? *Journal of Environmental Engineering and Landscape Management, 25*(2), 105–113.

Gámiz, B., Cox, L., Hermosín, M. C., Spokas, K., & Celis, R., (2017). Assessing the effect of organoclays and biochar on the fate of abscisic acid in soil. *Journal of Agricultural and Food Chemistry, 65*(1), 29–38.

Gao, X., & Wu, H., (2014). Aerodynamic properties of biochar particles: Effect of grinding and implications. *Environmental Science & Technology Letters, 1*(1), 60–64.

Githinji, L., (2014). Effect of biochar application rate on soil physical and hydraulic properties of a sandy loam. *Archives of Agronomy and Soil Science, 60*(4), 457–470.

Godfray, H. C. J., Beddington, J. R., Crute, I. R., Haddad, L., Lawrence, D., Muir, J. F., & Toulmin, C., (2010). Food security: The challenge of feeding 9 billion people. *Science, 327*(5967), 812–818.

Graber, E. R., Meller, H. Y., Kolton, M., Cytryn, E., Silber, A., Rav, D. D., & Elad, Y., (2010). Biochar impact on development and productivity of pepper and tomato grown in fertigated soilless media. *Plant and Soil, 337*(1), 481–496.

Guo, Y., Li, Y., Zhu, T., & Ye, M., (2013). Effects of concentration and adsorption product on the adsorption of SO_2 and NO on activated carbon. *Energy & Fuels, 27*(1), 360–366.

Hailegnaw, N. S., Mercl, F., Pračke, K., Száková, J., & Tlustoš, P., (2019). High temperature-produced biochar can be efficient in nitrate loss prevention and carbon sequestration. *Geoderma, 338*, 48–55.

Harter, J., Krause, H. M., Schuettler, S., Ruser, R., Fromme, M., Scholten, T., & Behrens, S., (2014). Linking N_2O emissions from biochar-amended soil to the structure and function of the N-cycling microbial community. *The ISME Journal, 8*(3), 660–674.

Herath, H. M. S. K., Camps-Arbestain, M., & Hedley, M., (2013). Effect of biochar on soil physical properties in two contrasting soils: An alfisol and an andisol. *Geoderma, 209*, 188–197.

Hou, D., (2020). *Sustainable Remediation of Contaminated Soil and Groundwater: Materials, Processes, and Assessment* (p. 10). Cambridge, MA: Butterworth- Heinemann.

Hou, D., Bolan, N. S., Tsang, D. C., Kirkham, M. B., & O'Connor, D., (2020). Sustainable soil use and management: An interdisciplinary and systematic approach. *Science of the Total Environment, 729*, 138961.

Husen, A., & Siddiqi, K. S., (2014). Carbon and fullerene nanomaterials in plant system. *Journal of Nanobiotechnology, 12*(1), 1–10.

Hussain, M., Farooq, M., Nawaz, A., Al-Sadi, A. M., Solaiman, Z. M., Alghamdi, S. S., & Siddique, K. H., (2017). Biochar for crop production: Potential benefits and risks. *Journal of Soils and Sediments, 17*(3), 685–716.

Jeffery, S., Verheijen, F. G., Van, D. V. M., & Bastos, A. C., (2011). A quantitative review of the effects of biochar application to soils on crop productivity using meta-analysis. *Agriculture, Ecosystems & Environment, 144*(1), 175–187.

Jiang, C., Bo, J., Xiao, X., Zhang, S., Wang, Z., Yan, G., & He, H., (2020). Converting waste lignin into nano-biochar as a renewable substitute of carbon black for reinforcing styrene-butadiene rubber. *Waste Management, 102*, 732–742.

Jones, D. L., Rousk, J., Edwards-Jones, G., DeLuca, T. H., & Murphy, D. V., (2012). Biochar-mediated changes in soil quality and plant growth in a three-year field trial. *Soil Biology and Biochemistry, 45*, 113–124.

Kammann, C., Ratering, S., Eckhard, C., & Müller, C., (2012). Biochar and hydrochar effects on greenhouse gas (carbon dioxide, nitrous oxide, and methane) fluxes from soils. *Journal of Environmental Quality, 41*(4), 1052–1066.

Karami, N., Clemente, R., Moreno-Jiménez, E., Lepp, N. W., & Beesley, L., (2011). Efficiency of green waste compost and biochar soil amendments for reducing lead and copper mobility and uptake to ryegrass. *Journal of Hazardous Materials, 191*(1–3), 41–48.

Kätterer, T., Roobroeck, D., Andrén, O., Kimutai, G., Karltun, E., Kirchmann, H., & De Nowina, K. R., (2019). Biochar addition persistently increased soil fertility and yields in maize-soybean rotations over 10 years in sub-humid regions of Kenya. *Field Crops Research, 235*, 18–26.

Khare, P., (2021). A comprehensive evaluation of inherent properties and applications of nano-biochar prepared from different methods and feedstocks. *Journal of Cleaner Production, 320*, 128759.

Kim, H. B., Kim, J. G., Kim, T., Alessi, D. S., & Baek, K., (2020). Mobility of arsenic in soil amended with biochar derived from biomass with different lignin contents: Relationships between lignin content and dissolved organic matter leaching. *Chemical Engineering Journal, 393*, 124687.

Kong, H., Zhang, Y., Li, Y., Cui, Z., Xia, K., Sun, Y., & Zhu, Y., (2013). Size-dependent cytotoxicity of nanocarbon blacks. *International Journal of Molecular Sciences, 14*(11), 22529–22543.

Kumar, H., Cai, W., Lai, J., Chen, P., Ganesan, S. P., Bordoloi, S., & Mei, G., (2020). Influence of in-house produced biochars on cracks and retained water during drying-wetting cycles: Comparison between conventional plant, animal, and nano-biochars. *Journal of Soils and Sediments, 20*(4), 1983–1996.

Kumar, M., Xiong, X., Wan, Z., Sun, Y., Tsang, D. C., Gupta, J., & Ok, Y. S., (2020). Ball milling as a mechanochemical technology for fabrication of novel biochar nanomaterials. *Bioresource Technology, 312*, 123613.

Kuzyakov, Y., Subbotina, I., Chen, H., Bogomolova, I., & Xu, X., (2009). Black carbon decomposition and incorporation into soil microbial biomass estimated by 14C labeling. *Soil Biology and Biochemistry, 41*(2), 210–219.

Law, P. N. K., Rahman, M. R., Hamdan, S., Jayamani, E., Bakri, M. K. B., & Sanaullah, K., (2020). Synthesis and characterization of micro-nano carbon filler from jatropha seeds. *BioResources, 15*(2), 3237–3251.

Lehmann, J., (2007). A handful of carbon. *Nature, 447*(7141), 143–144.

Lehmann, J., Rillig, M. C., Thies, J., Masiello, C. A., Hockaday, W. C., & Crowley, D., (2011). Biochar effects on soil biota–a review. *Soil Biology and Biochemistry, 43*(9), 1812–1836.

Li, B., Zhang, Q., Chen, Y., Su, Y., Sun, S., & Chen, G., (2021). Different crop rotation systems change the rhizosphere bacterial community structure of *Astragalus membranaceus* (Fisch) Bge. var. mongholicus (Bge.) Hsiao. *Applied Soil Ecology, 166*, 104003.

Lian, F., & Xing, B., (2017). Black carbon (biochar) in water/soil environments: Molecular structure, sorption, stability, and potential risk. *Environmental Science & Technology, 51*(23), 13517–13532.

Lian, F., Yu, W., Wang, Z., & Xing, B., (2018). New insights into black carbon nanoparticle-induced dispersibility of goethite colloids and configuration-dependent sorption for phenanthrene. *Environmental Science & Technology, 53*(2), 661–670.

Lian, F., Yu, W., Zhou, Q., Gu, S., Wang, Z., & Xing, B., (2020). Size matters: Nano-biochar triggers decomposition and transformation inhibition of antibiotic resistance genes in aqueous environments. *Environmental Science & Technology, 54*(14), 8821–8829.

Liao, H., Li, Y., & Yao, H., (2019). Biochar amendment stimulates utilization of plant-derived carbon by soil bacteria in an intercropping system. *Frontiers in Microbiology, 10*, 1361.

Liao, K. W., Guo, J. J., Fan, J. C., Lee, C. C., & Huang, C. L., (2019). Evaluation of rainfall kinetic energy and erosivity in northern Taiwan using kriging with climate characteristics. *Soil Use and Management, 35*(4), 630–642.

Libutti, A., Cammerino, A. R. B., Francavilla, M., & Monteleone, M., (2019). Soil amendment with biochar affects water drainage and nutrient losses by leaching: Experimental evidence under field-grown conditions. *Agronomy, 9*(11), 758.

Libutti, A., Francavilla, M., & Monteleone, M., (2021). Hydrological properties of a clay loam soil as affected by biochar application in a pot experiment. *Agronomy, 11*(3), 489.

Libutti, A., Mucci, M., Francavilla, M., & Monteleone, M., (2016). Effect of biochar amendment on nitrate retention in a silty clay loam soil. *Italian Journal of Agronomy, 11*(4), 273–276.

Liu, G., Zheng, H., Jiang, Z., & Wang, Z., (2018). Effects of biochar input on the properties of soil nanoparticles and dispersion/sedimentation of natural mineral nanoparticles in aqueous phase. *Science of the Total Environment, 634*, 595–605.

Liu, G., Zheng, H., Jiang, Z., Zhao, J., Wang, Z., Pan, B., & Xing, B., (2018). Formation and physicochemical characteristics of nano biochar: Insight into chemical and colloidal stability. *Environmental Science & Technology, 52*(18), 10369–10379.

Liu, W., Li, Y., Feng, Y., Qiao, J., Zhao, H., Xie, J., & Liang, S., (2020). The effectiveness of nanobiochar for reducing phytotoxicity and improving soil remediation in cadmium-contaminated soil. *Scientific Reports, 10*(1), 1–10.

Long, X., Luo, Y. H., Zhang, Z., Zheng, C., Zeng, C., Bi, Y., & Westerhoff, P., (2020). The nature and oxidative reactivity of urban magnetic nanoparticle dust provide new insights into potential neurotoxicity studies. *Environmental Science & Technology, 54*(17), 10599–10609.

Lu, K., Yang, X., Gielen, G., Bolan, N., Ok, Y. S., Niazi, N. K., & Wang, H., (2017). Effect of bamboo and rice straw biochars on the mobility and redistribution of heavy metals (Cd, Cu, Pb and Zn) in contaminated soil. *Journal of Environmental Management, 186*, 285–292.

Lyu, H., Gao, B., He, F., Zimmerman, A. R., Ding, C., Tang, J., & Crittenden, J. C., (2018). Experimental and modeling investigations of ball-milled biochar for the removal of aqueous methylene blue. *Chemical Engineering Journal, 335*, 110–119.

Ma, L., Zhong, H., & Wu, Y. G., (2015). Effects of metal-soil contact time on the extraction of mercury from soils. *Bulletin of Environmental Contamination and Toxicology, 94*(3), 399–406.

Ma, S., Jing, F., Sohi, S. P., & Chen, J., (2019). New insights into contrasting mechanisms for PAE adsorption on millimeter, micron-and nano-scale biochar. *Environmental Science and Pollution Research, 26*(18), 18636–18650.

Mahmoud, E., El Baroudy, A., Ali, N., & Sleem, M., (2020). Spectroscopic studies on the phosphorus adsorption in salt-affected soils with or without nano-biochar additions. *Environmental Research, 184*, 109277.

Makshut, N. A., Ngaini, Z., Wahi, R., Hussain, H., Mahmut, N. I., & Bahrin, N. Q., (2020). Nano-sized adsorbent from pyrolyzed sago activated sludge for removal of Pb (II) from aqueous solution. *Pertanika Journal of Science & Technology, 28*(3).

Meier, S., Curaqueo, G., Khan, N., Bolan, N., Rilling, J., Vidal, C., & Borie, F., (2017). Effects of biochar on copper immobilization and soil microbial communities in a metal-contaminated soil. *Journal of Soils and Sediments, 17*(5), 1237–1250.

Meyer, S., Glaser, B., & Quicker, P., (2011). Technical, economical, and climate-related aspects of biochar production technologies: A literature review. *Environmental Science & Technology, 45*(22), 9473–9483.

Mierzwa-Hersztek, M., Wolny-Koładka, K., Gondek, K., Gałązka, A., & Gawryjołek, K., (2020). Effect of coapplication of biochar and nutrients on microbiocenotic composition, dehydrogenase activity index and chemical properties of sandy soil. *Waste and Biomass Valorization, 11*(8), 3911–3923.

Moebius-Clune, B. N., (2016). *Comprehensive Assessment of Soil Health: The Cornell Framework Manual.* Cornell University.

Muhammad, N., Dai, Z., Xiao, K., Meng, J., Brookes, P. C., Liu, X., & Xu, J., (2014). Changes in microbial community structure due to biochars generated from different feedstocks and their relationships with soil chemical properties. *Geoderma, 226*, 270–278.

Mukherjee, A., & Lal, R., (2013). Biochar impacts on soil physical properties and greenhouse gas emissions. *Agronomy, 3*(2), 313–339.

Mukherjee, I., Das, S. K., Kumar, A., & Shukla, L., (2020). Sludge amendment affect the persistence, carbon mineralization and enzyme activity of atrazine and bifenthrin. *Bulletin of Environmental Contamination and Toxicology, 105*(2), 291–298.

Murtaza, G., Ahmed, Z., Usman, M., Tariq, W., Ullah, Z., Shareef, M., & Ditta, A., (2021). Biochar induced modifications in soil properties and its impacts on crop growth and production. *Journal of Plant Nutrition, 44*(11), 1677–1691.

Natural Resources Conservation Services: Soil Health, (2012). Retrieved from: http://www.nrcs.usda.gov/wps/portal/nrcs/main/soils/health/ The Soil Renaissance accepted this definition in 2014 (accessed on 25 April 2024).

Nguyen, T. T. N., Xu, C. Y., Tahmasbian, I., Che, R., Xu, Z., Zhou, X., & Bai, S. H., (2017). Effects of biochar on soil available inorganic nitrogen: A review and meta-analysis. *Geoderma, 288*, 79–96.

Oleszczuk, P., Ćwikła-Bundyra, W., Bogusz, A., Skwarek, E., & Ok, Y. S., (2016). Characterization of nanoparticles of biochars from different biomass. *Journal of Analytical and Applied Pyrolysis, 121*, 165–172.

Pascoli, M., Jacques, M. T., Agarrayua, D. A., Avila, D. S., Lima, R., & Fraceto, L. F., (2019). Neem oil based nanopesticide as an environmentally-friendly formulation for applications in sustainable agriculture: An ecotoxicological perspective. *Science of the Total Environment, 677*, 57–67.

Patriche, C. V., (2019). Quantitative assessment of rill and interrill soil erosion in Romania. *Soil Use and Management, 35*(2), 257–272.

Pérez-Esteban, J., Escolástico, C., Moliner, A., Masaguer, A., & Ruiz-Fernández, J., (2014). Phytostabilization of metals in mine soils using *Brassica juncea* in combination with organic amendments. *Plant and Soil, 377*(1), 97–109.

Qin, Y., Li, G., Gao, Y., Zhang, L., Ok, Y. S., & An, T., (2018). Persistent free radicals in carbon-based materials on transformation of refractory organic contaminants (ROCs) in water: A critical review. *Water Research, 137*, 130–143.

Rajput, V. D., Minkina, T., Ahmed, B., Singh, V. K., Mandzhieva, S., Sushkova, S., & Wang, B., (2022). Nano-biochar: A novel solution for sustainable agriculture and environmental remediation. *Environmental Research, 210*, 112891.

Ramezanzadeh, H., Reyhanitabar, A., Oustan, S., Mohammadi, M. H., & Van, D. Z. S. E. A. T. M., (2021). Enhanced sorption of cadmium by using biochar nanoparticles from ball milling in a sandy soil. *Eurasian Soil Science, 54*(2), 201–211.

Rehman, I., Riaz, M., Ali, S., Arif, M. S., Ali, S., Alyemeni, M. N., & Alsahli, A. A., (2021). Evaluating the effects of biochar with farmyard manure under optimal mineral fertilizing on tomato growth, soil organic c and biochemical quality in a low fertility soil. *Sustainability, 13*(5), 2652.

Sadegh-Zadeh, F., Tolekolai, S. F., Bahmanyar, M. A., & Emadi, M., (2018). Application of biochar and compost for enhancement of rice (*Oryza sativa* L.) grain yield in calcareous sandy soil. *Communications in Soil Science and Plant Analysis, 49*(5), 552–566.

Sahu, D., Kannan, G. M., & Vijayaraghavan, R., (2014). Carbon black particle exhibits size dependent toxicity in human monocytes. *International Journal of Inflammation, 2014.*

Sajjadi, B., Broome, J. W., Chen, W. Y., Mattern, D. L., Egiebor, N. O., Hammer, N., & Smith, C. L., (2019). Urea functionalization of ultrasound-treated biochar: A feasible strategy for enhancing heavy metal adsorption capacity. *Ultrasonics Sonochemistry, 51*, 20–30.

Sarfraz, R., Hussain, A., Sabir, A., Ben, F. I., Ditta, A., & Xing, S., (2019). Role of biochar and plant growth promoting rhizobacteria to enhance soil carbon sequestration—A review. *Environmental Monitoring and Assessment, 191*(4), 1–13.

Shao, L., Hou, C., Geng, C., Liu, J., Hu, Y., Wang, J., & BéruBé, K., (2016). The oxidative potential of PM10 from coal, briquettes and wood charcoal burnt in an experimental domestic stove. *Atmospheric Environment, 127*, 372–381.

Shen, Y., Tang, H., Wu, W., Shang, H., Zhang, D., Zhan, X., & Xing, B., (2020). Role of nano-biochar in attenuating the allelopathic effect from *Imperata cylindrica* on rice seedlings. *Environmental Science: Nano, 7*(1), 116–126.

Siedt, M., Schäffer, A., Smith, K. E., Nabel, M., Roß-Nickoll, M., & Van, D. J. T., (2021). Comparing straw, compost, and biochar regarding their suitability as agricultural soil amendments to affect soil structure, nutrient leaching, microbial communities, and the fate of pesticides. *Science of the Total Environment, 751*, 141607.

Sigmund, G., Huber, D., Bucheli, T. D., Baumann, M., Borth, N., Guebitz, G. M., & Hofmann, T., (2017). Cytotoxicity of biochar: A workplace safety concern? *Environmental Science & Technology Letters, 4*(9), 362–366.

Song, B., Cao, X., Gao, W., Aziz, S., Gao, S., Lam, C. H., & Lin, R., (2022). Preparation of nano-biochar from conventional biorefineries for high-value applications. *Renewable and Sustainable Energy Reviews, 157*, 112057.

Song, B., Chen, M., Zhao, L., Qiu, H., & Cao, X., (2019). Physicochemical property and colloidal stability of micron-and nano-particle biochar derived from a variety of feedstock sources. *Science of the Total Environment, 661*, 685–695.

Spokas, K. A., Novak, J. M., Masiello, C. A., Johnson, M. G., Colosky, E. C., Ippolito, J. A., & Trigo, C., (2014). Physical disintegration of biochar: An overlooked process. *Environmental Science & Technology Letters, 1*(8), 326–332.

Tan, S., Zhou, B., & Wang, Q., (2016). Effects of nanocarbon on the hydraulic parameters and the solute transport process for disturbed loessial soil. *Arabian Journal of Geosciences*, *9*(1), 1–13.

Tan, Z., Lin, C. S., Ji, X., & Rainey, T. J., (2017). Returning biochar to fields: A review. *Applied Soil Ecology*, *116*, 1–11.

Taner, S., Pekey, B., & Pekey, H., (2013). Fine particulate matter in the indoor air of barbeque restaurants: Elemental compositions, sources and health risks. *Science of the Total Environment*, *454*, 79–87.

Tao, H., Pan, W. L., Carter, P., & Wang, K., (2019). Addition of lignin to lime materials for expedited pH increase and improved vertical mobility of lime in no-till soils. *Soil Use and Management*, *35*(2), 314–322.

Tassinari, D., Andrade, M. L. D. C., Dias, Jr. M. D. S., Martins, R. P., Rocha, W. W., Pais, P. S. A. M., & De Souza, Z. R., (2019). Soil compaction caused by harvesting, skidding and wood processing in eucalyptus forests on coarse-textured tropical soils. *Soil Use and Management*, *35*(3), 400–411.

Wan, B., Hou, J., & Guo, L. H., (2017). Safety of carbon nanotubes. In: *Industrial Applications of Carbon Nanotubes* (pp. 405–431). Elsevier.

Wang, D., Zhang, W., Hao, X., & Zhou, D., (2013). Transport of biochar particles in saturated granular media: Effects of pyrolysis temperature and particle size. *Environmental Science & Technology*, *47*(2), 821–828.

Wang, L., Wu, W. M., Bolan, N. S., Tsang, D. C., Li, Y., Qin, M., & Hou, D., (2021). Environmental fate, toxicity and risk management strategies of nanoplastics in the environment: Current status and future perspectives. *Journal of Hazardous Materials*, *401*, 123415.

Wang, M., Zhu, Y., Cheng, L., Andserson, B., Zhao, X., Wang, D., & Ding, A., (2018). Review on utilization of biochar for metal-contaminated soil and sediment remediation. *Journal of Environmental Sciences*, *63*, 156–173.

Wang, S. L., & Nguyen, A. D., (2018). Effects of Zn/B nanofertilizer on biophysical characteristics and growth of coffee seedlings in a greenhouse. *Research on Chemical Intermediates*, *44*(8), 4889–4901.

Wang, Y. Y., Jing, X. R., Li, L. L., Liu, W. J., Tong, Z. H., & Jiang, H., (2017). Biotoxicity evaluations of three typical biochars using a simulated system of fast pyrolytic biochar extracts on organisms of three kingdoms. *Acs Sustainable Chemistry & Engineering*, *5*(1), 481–488.

Wang, Y., Wei, Y., & Sun, J., (2016). Biochar application promotes growth parameters of soybean and reduces the growth difference. *Communications in Soil Science and Plant Analysis*, *47*(12), 1493–1502.

Watzinger, A., Feichtmair, S., Kitzler, B., Zehetner, F., Kloss, S., Wimmer, B., & Soja, G., (2014). Soil microbial communities responded to biochar application in temperate soils and slowly metabolized 13C-labelled biochar as revealed by 13C PLFA analyses: Results from a short-term incubation and pot experiment. *European Journal of Soil Science*, *65*(1), 40–51.

Weber, K., & Quicker, P., (2018). Properties of biochar. *Fuel*, *217*, 240–261.

Woolf, D., Amonette, J. E., Street-Perrott, F. A., Lehmann, J., & Joseph, S., (2010). Sustainable biochar to mitigate global climate change. *Nature Communications*, *1*(1), 1–9.

Wu, P., Ata-Ul-Karim, S. T., Singh, B. P., Wang, H., Wu, T., Liu, C., & Chen, W., (2019). A scientometric review of biochar research in the past 20 years (1998–2018). *Biochar*, *1*(1), 23–43.

Xiao, X., & Chen, B., (2017). A direct observation of the fine aromatic clusters and molecular structures of biochars. *Environmental Science & Technology, 51*(10), 5473–5482.

Yang, X., Ng, W., Wong, B. S. E., Baeg, G. H., Wang, C. H., & Ok, Y. S., (2019). Characterization and ecotoxicological investigation of biochar produced via slow pyrolysis: Effect of feedstock composition and pyrolysis conditions. *Journal of Hazardous Materials, 365*, 178–185.

Yang, Y., Zhou, B., Hu, Z., & Lin, H., (2020). The effects of nano-biochar on maize growth in northern Shaanxi province on the loess plateau. *Appl. Ecol. Environ Res., 18*, 2863–2877.

Yao, X., Ji, L., Guo, J., Ge, S., Lu, W., Cai, L., & Zhang, H., (2020). Magnetic activated biochar nanocomposites derived from wakame and its application in methylene blue adsorption. *Bioresource Technology, 302*, 122842.

Yue, L., Lian, F., Han, Y., Bao, Q., Wang, Z., & Xing, B., (2019). The effect of biochar nanoparticles on rice plant growth and the uptake of heavy metals: Implications for agronomic benefits and potential risk. *Science of the Total Environment, 656*, 9–18.

Zama, E. F., Reid, B. J., Arp, H. P. H., Sun, G. X., Yuan, H. Y., & Zhu, Y. G., (2018). Advances in research on the use of biochar in soil for remediation: A review. *Journal of Soils and Sediments, 18*(7), 2433–2450.

Zhang, K., Mao, J., & Chen, B., (2019). Reconsideration of heterostructures of biochars: Morphology, particle size, elemental composition, reactivity and toxicity. *Environmental Pollution, 254*, 113017.

Zhang, Y., Zhao, G., Xuan, Y., Gan, L., & Pan, M., (2021). Enhanced photocatalytic performance for phenol degradation using ZnO modified with nano-biochar derived from cellulose nanocrystals. *Cellulose, 28*(2), 991–1009.

Zhi, L., Sun, H., Xu, L., & Cai, Y., (2020). Distribution and elimination of trifluoropropylmethylsiloxane oligomers in both biosolid-amended soils and earthworms. *Environmental Science & Technology, 55*(2), 985–993.

Zhou, B. B., Chen, X. P., & Henry, L., (2020). The effect of nano-biochar on soil, water, and nutrient loss of a sloping land with different vegetation covers on Loess plateau of China. *Applied Ecology and Environmental Research, 18*(2), 2845–2861.

CHAPTER 9

Nano-Composites and Nano-Biochar: A Novel Solution to Alleviate Heavy Metal Stress in Crops

DEBOJYOTI MOULICK,[1] ARKABANEE MUKHERJEE,[2] SUBRATA MAHANTA,[3] BHABEN CHOWDHARA,[4] BINAYA KUMAR PATTNAIK,[5] ANOOP KUMAR MUKHOPADHYAY,[6] NIHAL GUJRE,[2] JAYA SHREE,[3] SUBHAS CHANDRA SANTRA,[1] and SWATI HAZRA[7]

[1]*Department of Environmental Science, University of Kalyani, Nadia, West Bengal, India*

[2]*Indian Institute of Tropical Meteorology, Panchawati, Pashan, Pune, Maharashtra, India*

[3]*Department of Chemistry, NIT Jamshedpur, Adityapur, Jamshedpur, Jharkhand, India*

[4]*Department of Botany, Faculty of Science and Technology, Arunachal University of Studies, Knowledge City, Namsai, Arunachal Pradesh, India*

[5]*Symbiosis Institute of Geoinformatics, Pune, Maharashtra, India*

[6]*Department of Physics, Sharda School of Basic Sciences and Research, Sharda University, Greater Noida, Uttar Pradesh, India*

[7]*Sharda School of Agricultural Sciences, Sharda University, Greater Noida, Uttar Pradesh, India*

Nanomaterials and Nano-Biochar in Reducing Soil Stress: An Integrated Approach to Sustainable Agriculture. Vishnu D. Rajput, & Tatiana Minkina (Eds.)

ABSTRACT

Heavy metal (HM) toxicity inhibits the normal metabolic processes of living organisms and can be lethal to both plants and animals. Therefore, the presence of HMs in ecosystems and their entry into the food chain is one of the key concerns of researchers worldwide. Anthropogenic activities related to agriculture, industries, and sewage disposal cause the accumulation of HMs in agricultural soil and irrigation water. These HMs can have detrimental effects on the metabolism and growth of plants, thereby reducing crop yield. Nanotechnology holds a promising future in the field of sustainable agriculture by not only conserving resources but also promoting tolerance to stress conditions, including those caused by HMs. In this chapter, we provide a concise account of the effects of HMs on global food security and the potential role of nanotechnology, especially through nano-composites (NCs) and nano-biochar (NB), in alleviating HM stress. A thorough yet succinct overview of NCs and NB, including their characterization and function, is also presented.

9.1 INTRODUCTION

In the recent decade, climate change has severely impacted agriculture, which is further compounded by industrialization, urbanization, and extensive agricultural practices (Choudhury & Moulick, 2022). Today, the global population has crossed the 7+ billion mark, which is anticipated to reach 9.6 billion by 2050 (UN, 2022). Therefore, we will need an additional 1 billion tons of cereals by that time to ensure food security (FAO, 2022). According to the United Nations' 2030 Agenda in relation to achieving sustainable development, 17 targets, generally called Sustainable Development Goals (SDGs), were adopted. Among these 17 prime areas, the focus was set on the eradication of hunger and poverty, providing a healthcare system along with hygiene, and holistic natural resource utilization with the intention to ensure a suitable and sustainable socio-economic-environment for the world population and future generations. The goal of zero hunger, termed as SDG 2, has to be attained by 'end hunger, achieve food security and improved nutrition and promote sustainable agriculture' (UN, 2015). Several authors (Pradhan, 2019; Pradhan et al., 2017) have pointed out the inter- and trans-dependency of the 17 prime areas along with their multi-layered nature; thus, both antagonistic as well as synergistic interactions among them are required to be addressed in an integrated and all-inclusive manner to attain the SDGs. According to the reports of FAO (2019) and Development Initiatives (2017), the current world situation is quite shocking, with 2 billion people suffering from obesity (overweight) whereas another 2 billion people are deprived of essential micronutrients (iron, vitamin A) and a total of 821 million people are reported to be undernourished. Several authors attributed the importance of the agricultural sector and its adverse impacts on climate by contributing significantly towards greenhouse gas (GHG) emissions (10–12%) and depletion of groundwater level (70% of freshwater withdrawals) which is mostly used for irrigation and other operations (Smith et al., 2014; Foley et al., 2011). Nevertheless, Alexander et al. (2017); and Gustavsson et al. (2011) were of the opinion that wastage of food products, as much as $1/3^{rd}$, poses significant obstacles in achieving SDG 2 goals.

In field conditions, crops usually experience a diverse range of stressors of different intensities. Any agro-environmental component (biotic and/or abiotic) is considered a constraint or stressor when present either in excess or below the optimal level. These stressors may include weed species exerting adverse impacts by limiting essential resources to key crops (Ghosh et al., 2020a, b, 2021, 2022); the presence of excess CO_2 and O_3 (Mukherjee & Hazra, 2022; Hazra et al., 2019); sudden waterlogging that creates submergence stress (Panda et al., 2022; Nayak et al., 2022); limited phyto-availability of groundwater (Jadhav et al., 2022; Sagar et al., 2022), etc. Thus, the challenge in front of researchers is to ensure a steady crop yield by opting for suitable means at the field level (Roy et al., 2022; Maitra et al., 2022).

Owing to the huge economic advances and rapid growth in various fields, such as industry and agriculture, the environment is steadily becoming more unhygienic with each passing day (Ali et al., 2019; Moulick et al., 2021). Environmental pollutants are either of natural origin or enter the ecosystem due to anthropogenic (human) activity. Mining operations, unwise biomass burning practices, random discharge of industrial waste, and dumping processes are a few examples of such anthropogenic activities responsible for the buildup of pollutants in the environment (Bhunia et al., 2017). Pollutants that exhibit toxicity upon crossing the MRL (maximum residue limits) standard have received sharp focus from investigators (Tóth et al., 2016; Zhang et al., 2019). Among the HMs, lead (Pb), arsenic (As), cadmium (Cd), copper (Cu), and zinc (Zn) are responsible for a frightening situation associated with agro-environmental and public health issues (Zhang et al., 2019; Su et al., 2014). Apart from their adverse effects on crop life cycles, a considerable accumulation of HMs in consumable portions and their subsequent transmission into the food web have become obstacles to global food security. The contamination of agro-ecosystems with HMs demands drastic and practical solutions to minimize the hazards as soon as possible. Observations made by Xiao et al. (2017) suggested that arable lands located close to electroplating and cement factories had a significantly greater amount of HM content.

Biochar can be best defined as a carbon-enriched substance obtained after the pyrolysis of biomass (Yadav et al., 2020) that is mainly utilized as a fertility/productivity enhancer for arable soil, as well as used for the sequestration of atmospheric carbon (Spokas et al., 2009; Yadav et al., 2019). Due to its porous characteristics with greater surface functionality, biochar has been successfully utilized as an effective photocatalyst, agent for immobilization of various enzymes, and absorbent for purification by eliminating inorganic and organic pollutants, and substances restricting the availability to crops of toxic elements in soil (Beesley et al., 2011; Kookana et al., 2011; Yang et al., 2013).

Various origins of biomass, such as feedstocks, husk, etc., and a wide range of pyrolysis conditions have been considered for biochar making, leading to the creation of a wide range of biochar, which in turn can modulate the interactive behavior, i.e., mechanisms among the HMs and biochar. Having a huge porous surface area with numerous functional groups, enriched mineral content, ionic diversity, loaded with organic substances makes the biochar an ideal candidate to entrap a wide range of HMs. The entrapment of numerous HMs by biochar can be attributed to phenomena such as the conversion of inorganic HMs into organic HMs, facilitation of electrostatic attraction, formation of complexes, anionic/cationic exchange, or even precipitations (Lahori et al., 2017; Xu et al., 2018; Wang et al., 2018).

Minimizing the particle size of biochar (BC) to its nano-/micro-scale is considered a competent technique that can enhance the characteristic features like specific surface area (SSA), porosity, flexibility, and other associated attributes of BC. Apart from these, nano-biochar (N-BC) elevates the sub-optimal performance of BC to a customized high-quality NP, called nano-biochar (Weber & Quicker, 2018). On the other hand, as compared to BC, N-BC usually has a smaller size (<100.00nm), which enables them with greater mobility and transmission in the soil-water system (Chen et al., 2017; Wang et al., 2013). Moreover, several authors were of the opinion that as a carrier, N-BC could smoothen the passage of regular and/or advantageous solutes and contaminants as compared to BC, having only positive effects of retention of nutrients and entrapment of hazardous chemicals (Ahmad et al., 2014; Lian & Xing, 2017). Chen et al. (2018) reported that BC nanoparticles (NPs) unpleasantly enhanced P leakage in alkaline soil. Kim et al. (2018) also found the augmentation of desorption and movement of As in soil that was treated with BC fraction less than 0.45 μm. Moreover, the mean sizes of BC having a breadth of 4–5 μm and length up to ~500 nm are generally coined as N-BC. Such variation in terms of particle size has provided an additional parameter to relate the efficacy of different N-BC to the SSA of N-BC particles, which is basically prejudiced by their sizes itself and thus demands to be considered in the near future (Song et al., 2022).

Keeping in mind the challenges of basic BC limiting their use in the agro-environmental sector, a mounting number of researchers have directed their focus to develop nano-composite materials based on BC. Huang et al. (2019) observed that composites based on BC have an edge over conventional BC in terms of physicochemical attributes and numerous possibilities of linking other materials with BCs. Composites based on BCs (C-BC) demonstrated improved performance due to their packing with organic materials/inorganic BC. Biochar-based composites can be categorized into (i) nano-metal oxide/hydroxide BC composites and (ii) BC-magnetic composites (Huang et al., 2019). In this chapter, our aim will be oriented towards presenting our assessment of the potentials of nano-composites and nano-biochar's applications, agro-environmental issues, and the future.

9.2 CONSEQUENCES OF HM CONTAMINATION ON CROPS AND THEIR PRODUCTIVITY

As seen by variations in their pigment level, amount of water, dry weight, and growth, various plants respond to nutrient and metal concentrations in different ways (Sridhar et al., 2011; Moulick et al., 2016b, 2018a, b, c; Saha et al., 2019). Indicators of soil pollution and the plant's physiological state can be detected in the unique light absorption and reflectance behavior that result from all these changes in plant features. Many reports mentioned that plant HMs as well as nutrient stress results in alterations in the spectral reflectivity of the crop species (Su et al., 2007; Sridhar et al., 2007, 2009, 2011). Numerous physiological and biochemical alterations in the plants may result from these modifications, which may affect how much or how little nutrient is present in the vegetables. HM poisoning in plants has physiological impacts on seed germination, causes significant decreases in respiration, transpiration, growth rates, adjustments in photosynthetic efficiency, abnormalities in nutritional homeostasis, as well as variations in the adoption rates of Mn, K, Mg, and Ca (Moulick et al., 2017, 2019; Ciscato et al., 1999). Manzoor et al. (2018) mentioned that *Raphanus* and *Phaseolus* exhibit distinct symptoms of leaves along with a decline in the root: shoot ratio and biomass ratio.

There have been reports of Cd treatments inhibiting plant growth, as well as signs of leaf chlorophyll degradation at 10M Cd and necrotic patches at 100M Cd, as well as browning of roots in both treatments (Millan et al., 2009). Phosphoenolpyruvate carboxylase activity, which is involved in the anaplerotic fixation of CO_2 into organic acids, elevated root extract content of Cd-exposed plants. Citrate synthase, isocitrate dehydrogenase, and malate dehydrogenase activities were also demonstrated to significantly increase in leaf extracts at 100M Cd, although fumarase activity was reduced. It is well recognized that metal toxicity alters cells at the cellular level by destroying membranes, obstructing electron transport, blocking or activating enzymes, along with nucleic acids, etc. (Chen et al., 2003). The development of oxidative stress and the replacement of elements like Zn, Fe, and Mn, which are vital cofactors for many enzymes, are two potential causes of these diseases. Numerous studies have proven the existence of a relationship between imposing oxidative stress and exposure to greater content of trace metals (Mazumder et al., 2022; Choudhury et al., 2021a, b, 2022a, b; Mazumder et al., 2021; Lin et al., 2007). According to McLaughlin et al. (1996), the effects of heavy metals (HMs) on plants include structural damage, growth inhibition, and a deterioration in plant biochemical and physiological processes. Some of the specific effects of heavy metal toxicity include disruption of transport systems for vital nutrients and ions, denaturation and deactivation of enzymes, blockage of polynucleotides, displacement and/or replacement of vital ions from cellular sites, rupturing of cell and organelle membrane permeability, and dismantling and dismissal of enzymes. Surplus-optimal concentrations of heavy metals such as cadmium, lead (Pb), mercury (Hg), copper (Cu), zinc (Zn), and nickel (Ni) have an impact on plant growth, development, and yield (Sresty & Rao, 1999).

In sludge-added, HM-contaminated soils, Pascual et al. (2010) discovered that pepper fruits from those plants had lower quantities of capsaicin and dihydrocapsaicin, resulting in less pungency. Sridhar et al. (2014) measured the elemental concentrations of soil, shoots, roots, and fruit samples. Higher levels of copper (Cu) and zinc (Zn) were found in the plant shoots of radish (*Raphanus raphanistrum* subsp. *sativus)*, lettuce (*Lactuca sativa*), pepper (*Piper nigrum*), tomato (*Solanum lycopersicum*), and collard (*Brassica oleracea*). Tomato plants' fruits were found to have much higher concentrations of Cu and Zn compared to other plants.

Changes in metal and nutrient intake, as well as changes brought on by HM pollution, are important factors to consider (Moulick et al., 2016a, 2022). These changes may be caused by variations in plant dry matter and root exudates released into the soil. The biogeochemical interactions of organic acids released in the soil through root exudates will also affect the availability of metals and nutrients for plant absorption as plants age and roots extend in the soil. This mechanism explains why pepper (*Piper nigrum*) and tomato (*Solanum lycopersicum*) plants digest Cu and Zn more quickly than other vegetable plants. According to Koo et al. (2013), Zn and Cu interact with the organic acids in root exudates to produce organometallic complexes, enhancing plant uptake. They discovered that all three vegetable crops were hazardous due to high Zn in the growing media. Young leaves exhibited chlorosis, coralloid roots turned brown, and plant growth was severely hampered. Shoot fresh weight (FW) diminished over time as Zn concentrations rose.

According to Xiong & Wang's (2005) observations, Cu has a negative impact on Chinese cabbage (*Brassica pekinensis*) seed germination. The median fatal dosage was 0.348 mmol L^{-1}, and the 0.5 mmol L^{-1} Cu treatment noticeably minimized the rate of germination. Additionally, Cu reduced the young seedlings' root and shoot lengths, whereas a very low Cu concentration (0.008 mmol L^{-1}) stimulated the shoots to elongate. Zhang & Zhou (2005) claim that the germination of vegetable seeds was harmful when aluminum coagulants were used at a specific dose. Due to the high Cu levels in the growth media, which also resulted in chlorosis in the new leaves and brown and stunted coralloid roots, vegetable crops were also harmed (Manzoor et al., 2018). According to Alia et al. (2015), the amount of HMs meaningfully dwindled spinach biomass. Biomass (both fresh and dry) decreased by 10.1 to 35.1% at the highest Cd dose. While the FW and DW of Pb-treated plants declined by 8% and 28%, respectively. The total protein, fiber, and moisture levels of spinach all decreased by 31%, 29%, and 33%, respectively, at the highest dose of Cd. The total protein, fiber, and moisture contents decreased by 23%, 22%, and 29%, respectively, upon exposure to Pb, compared to the control. Due to the high concentration of heavy metals, the amounts of sodium, potassium, calcium, iron, magnesium, manganese, and copper in spinach dropped. A high Pb level in mustard greens (*Brassica juncea*) demonstrated a 77% decrease in protein content. Lin et al. (2007) and Mahmood et al. (2005) found that the protein content of sunflower (*Helianthus annuus*) and wild carrot *(Daucus carota)* significantly decreased in higher concentrations of Cd. Similar results were also found in algae and Rapeseed (*Brassica napus*) under Zn treatments (John et al., 2009). Xu et al. (2006) attributed decline in protein amount to the acceleration of protein degradation with increasing protease activity or heavy metals' effects on nitrogen metabolism. Many research papers mentioned that the adverse impact of trace metals (Al, Cd, Fe, and As) on plant development and yield at different growth stages (Chowardhara et al., 2019a, b, 2020, Regon et al., 2021; Awasthi et al., 2022; Moulick et al., 2016b).

9.3 NANO-COMPOSITES AND NANO-BIOCHAR

Soil provides a medium and habitat for the growth of plants. Discharge of harmful chemicals pollutes the soil to a large extent. Contamination caused by HMs, for example, Hg, Cu, Pb, Cd, and As, must be removed because these HMs are highly poisonous and persist in the environment for a long time. Therefore, nanostructured biochar is used as an eco-friendly material for the remediation of HMs from the soil. Biochar, also known as prochar, is a carbonized solid that can be obtained from the thermal disintegration of agricultural wastes (Hossain et al., 2011). These nanomaterials have emerged with extraordinary properties that act as a sustainable solution for eliminating pollutants from contaminated soil. Nano-biochar contains a much higher surface area than macro-biochar, which can act as a very good adsorbent for removing toxic HMs from polluted soil. These nanostructured materials exhibit superior physicochemical properties, such as high catalytic activity, large surface area, unique nanostructure, etc., over bulk biochar (Song et al., 2022).

The application of nano-biochar can be seen in various other fields, such as electrochemistry (electrodes), biocatalysis, biorefinery, capacitors, sensors, photocatalytic materials, etc. Biochar can be obtained from wood, forestry waste, growing algae organisms, etc. The extraordinary properties of biochar, such as a good permeable structure, a greater surface area, and the presence of carbonyl and hydroxyl groups, help in the degradation and exclusion of HM contaminants from the soil. The most common technique for the production of biochar is mechanical grinding. Biochar nano-composites have graphitic features, negative zeta potential, and better stability than bulk biochar.

9.3.1 TYPES OF BIOCHAR NANO-COMPOSITES

9.3.1.1 OXIDE/HYDROXIDE BIOCHAR NANOPARTICLES

These nanoparticles can be obtained using three methods:

- Fortification of target element via bioaccumulation
- Pre-treatment of biomass using metal salt
- Pyrolysis insertion method of non-metal oxide

9.3.1.2 FUNCTIONAL BIOCHAR NANO-COMPOSITES

Biochar is produced by coating functional NPs such as graphene oxides, carbon nanotubes, chitosan, etc., which help in the removal of harmful contaminants from soil. The pollutant removal ability of nano-composites can also be enhanced by carrying out modifications using metal oxides to enhance sorption ability.

9.3.1.3 MAGNETIC BIOCHAR COMPOSITES

This biochar nano-composite enhances surface area, has good catalytic degradation capacity, and provides adequate availability of active sites that help improve the rate of biochar adsorption (Gámiz et al., 2017). When biochar is modified with iron oxides/salts, the recyclability of biochar is enhanced due to increased magnetic properties. This makes it easier to separate biochar from water using a regular magnet, which was difficult to carry out due to the very small size of biochar (Figure 9.1).

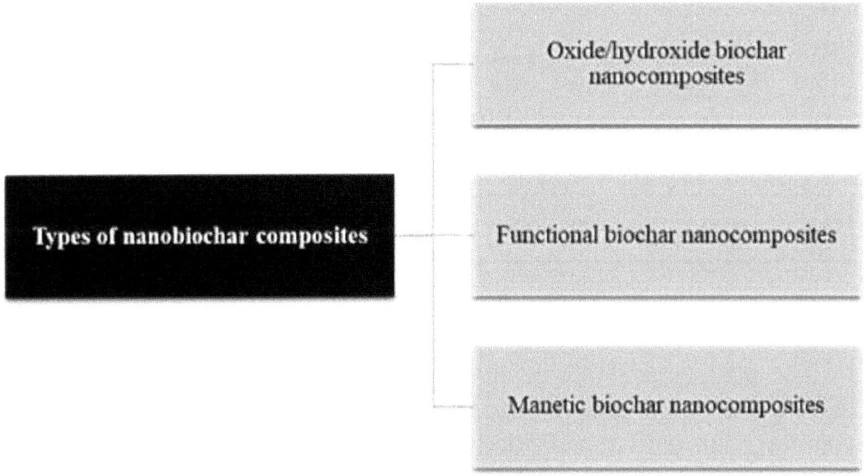

FIGURE 9.1 Classification of nano-biochar composites.

9.3.2 PREPARATION OF NANO-COMPOSITES AND NANO-BIOCHAR

Nano-composites and nano-biochar can be synthesized from biomass following different processes as compiled in Table 9.1.

TABLE 9.1 Preparation of Biochar and Nano-Biochar from Different Biomass Sources

Sources of biomass	Preparation Details (Temperature etc.)	C content	References
Jatropha Seed	Preparation method – Ball milling Time – 1,800 minutes, Number of balls – 25, Weight – 0.5 g, Particle size – 600 nm Biochar temperature – 180°C (microwave)	81.2 %	Law et al. (2020)
Pine white wood	Preparation method – Ball milling Speed – 575 rpm, Time – 100 minutes, Number of balls – 800, Ball size – 2.4 mm, Particle size – 60±10 nm, Biochar temperature – 525±1°C	83.1±2.5 %	Naghdi et al. (2017)
Wheat Straw	Preparation method – Sonication Particle size – 20–100 nm, Power – 60 kJ, Temperature – < 20°C, Amplitude – 80%, Probe size – 30 mm	27.88 %	Oleszczuk et al. (2016)
Sago effluent	Preparation method – Carbonization Particle size – 45–75 nm, Temperature – 500°C, pH – 7, Time – 90 minutes, Treatment – HCl	0.08 %	Makshut et al. (2020)
Waste lignin	Preparation method – Carbonization Particle Size – 473.6±81.9 nm, Temperature – 800°C, Treatment – Ball milling for 4 hours	84.44 %	Jiang et al. (2020)
Rice Straw	Preparation method – Centrifugation Biochar temperature – 400°C, Particle size – 1.9 nm, Speed – 10,000 rpm, Time – 30 minutes	NA	Lian et al. (2020)
Wood Chips	Preparation method – Centrifugation Particle size – <100 nm, Speed – 3,500 rpm, Time – 30 minutes	17.7 %	Song et al. (2019)

The main procedures involved in the production of nano-composites and nano-biochar include: size reduction, carbonization, sonication, and centrifugation and sedimentation.

9.3.2.1 SIZE REDUCTION OF BULK BIOCHAR

Biochar size reduction processes can increase the nanoparticle content (Liu et al., 2018). Many authors recommend the hand grinding method using a mortar to reduce the size of bulk biochar (Lonappan et al., 2016; Lian et al., 2020). Kumar et al. (2020) advocate the ball milling method to improve nano-biochar yield, including the reduction of particle size for large-scale manufacturing. Biochar is broken down into nano-size during the ball milling process by the contact of metallic balls. Operational parameters for ball milling comprise ball diameter, parent materials, ball to biochar mass ratio, milling duration, running speed, temperature, and liquid media (Lyu et al., 2018). Wet and dry techniques can be used for ball milling of biochar. In comparison to dry milling, the wet process provides more benefits such as more functional groups, smaller particles with greater dispersivity (Yuan et al., 2020). Xu et al. (2020) detected that solvent-assisted ball milling in wet milling showed an enhancement of 64% in biochar's surface area, which outperformed ball milling with salt assistance. Xiao et al. (2020) reduced particle size to less than 1 μm through the ball milling method using cow bone meal biomass. Munkhbayar et al. (2013) pointed out that the formation of dense agglomerates of nanoparticles can create challenges for achieving optimal ball milling efficiency. But prior to ball milling, if the biochar is subjected to preconditioning at –80°C, it can suppress agglomerate formation (Naghdi et al., 2017). Biochar blends require further dispersion in water or solvents after milling to produce a more uniform particle distribution prior to fractionation (Li et al., 2017). Although pre-conditioning at –80°C can reduce the likelihood of particle aggregation, the technique has been criticized as an expensive one that will drive up the price of producing nano-biochar and prevent it from being mass-produced. Mixing magnetic substances such as iron oxides in biochar can minimize particle agglomeration and maximize particle distribution, resulting in more effective water treatment (Li et al., 2020).

9.3.2.2 CARBONIZATION

Carbonization is the process of transforming waste biomass into carbon/energy-rich biomass with enhanced heating value (Amer & Elwardany, 2020). It can modify surface properties and the size of nano-biochar and get rid of infused salts. Jiang et al. (2020) utilized the carbonization process on waste lignin for the synthesis of nano-biochar. Carbonization of biomass can be achieved through methods involving various combinations with excess heat treatment along with the application of chemicals such as flash carbonization, hydrothermal carbonization, gasification, and pyrolysis (Meyer et al., 2011; Lin et al., 2016; Pang, 2019).

During pyrolysis, organic materials undergo thermal decomposition at a temperature ranging from 250 to 900°C in the absence of oxygen (Osayi et al., 2014). Biomass gasification for syngas, fast pyrolysis for bio-oil, and biocrude synthesis through hydrothermal treatment are examples of biochar in varied yields (Demirbas, 2004). The ligno-cellulosic ingredients, such as lignin, cellulose, hemicelluloses, go through reaction processes including fragmentation, depolymerization, and cross-linking at particular temperatures during the process, producing a variety of products in the form of solid (char), gas (carbon dioxide, carbon monoxide, hydrogen, syngas), and liquid (bio oil) (Yaashikaa et al., 2020). For the manufacture of biochar, a variety of reactor designs are employed, including agitated sand rotating kilns, wagon reactors, bubbling fluidized beds, and paddle kilns. Indigenous properties of feed stocked biomass influence the production of biochar during the pyrolysis process. Temperature is a detrimental factor that controls the efficiency of biochar produced (Wei et al., 2019). Based on the reaction time, the pyrolysis method can be classified as fast pyrolysis (approximately 1 second) and slow (minutes to days duration). The source of heat in these methods can be electrical heating, microwave, and fuel burning (Bridgwater, 2007). In fast pyrolysis, solid biomass is subjected to a fast rate of heating of about more than 100C per minute. Biomass particles interact with the pyrolysis fumes for 0.5 to 2 seconds. Fast pyrolysis technologies must reduce the residence time of fumes in the hot zone to produce good-quality bio-oil, which can be achieved by rapid extinguishing or cooling of fumes (Wang et al., 2014). On the contrary, slow pyrolysis requires a longer residence time of 1 hour with a reduced rate of heating of 5–7°C per minute (Liu et al., 2015). Jensen et al. (2000) pointed out a few limitations in the pyrolysis process, such as maintaining an appropriate reaction time along with high and controlled heat rate. The author also mentioned that liquid recovery and rapid removal of char in the fast pyrolysis process are challenging.

In hydrothermal carbonization, biomass is combined with water and placed in a sealed reactor under elevated pressure for several hours. During the process, gradual heating is used to maintain equilibrium (Meyer et al., 2011). This process yields different products based on the variation of temperature such as hydrothermal liquefaction or bio oil (250–400°C), hydrothermal carbonization or biochar (< 250°C), and hydrothermal gasification (> 400°C) (Khorram et al., 2016). Libra et al. (2011); and Funke & Ziegler (2010) raised concerns regarding the elasticity capacity of materials used under high pressure during hydrothermal carbonization. Additionally, post-treatment of char and heat recovery from high-temperature water is also a challenging task in this process.

During the gasification process, partial oxidation of biomass occurs at around 800°C in the gasification chamber, while the operation is conducted at atmospheric pressure or elevated pressure levels (Bridgwater et al., 2002; Bridgwater, 2007). In gasification, syngas is the major product, and char is produced in very small amounts as a byproduct. Shoot and aerosol formation, along with tar dehydration, clogging of particulate filters, and corrosion, are considered as limitations in the gasification process (Nunes et al., 2008).

In the flash carbonization process, 0.8 to 1.5 kg of air Kg^{-1} of biomass is blown towards the carbonization bed of biomass, and at the same time, a flash fire is ignited and forced upwards in the direction of the biomass bed with an elevation pressure of 1–2 MPa. Process parameters such as reaction duration are maintained under 30 minutes, meanwhile, the temperature in the reactor varies from 300 to 600°C (Meyer et al., 2011). Biomass gets converted into solid and liquid phase products (Yaashikaa et al., 2020). Upon ignition, under particular process conditions and with certain feedstocks, a rapid increase in pressure within the carbonization container can increase the risk during the process (Wade et al., 2006).

9.3.2.3 SONICATION

Biochar adsorption capacity can be enhanced through physical modification approaches such as the ultrasonic method. The high-energy ultrasound used in the sonication process causes the biochar suspended in the suspension to disintegrate into smaller pieces (Li et al., 2013). Shockwaves used in this technique cause the biochar's microporous area to grow by clearing clogged pores and exfoliating the carbon structure. The adsorption capacity of biochar increases due to the leaching of minerals into the pore spaces (Sajjadi et al., 2020). Sonication exfoliates the fine fragments of biochar, resulting in nano-biochar. Based on the amount of heterogeneous carbon in bulk biochar, an increase in the duration of sonication can result in a greater quantity of nano-biochar (Liu et al., 2018). The sonication process can remove blocked pore spaces in biochar and create a homogeneous surface (Shan et al., 2019). Process parameters like temperature, amplitude, power variation, and pore size during the sonication process can influence the characteristics of nano-biochar (Anupama & Khare, 2021). Wang et al. (2022) introduced chemical and microwave treatment coupled with the ultrasonic method for biochar preparation, resulting in enhanced surface area without high-temperature pyrolysis.

9.3.2.4 CENTRIFUGATION AND SEDIMENTATION

In order to isolate nano-biochar, sedimentation and membrane filtering are the two most common techniques. An identical sedimentation technique under static conditions to obtain nano-biochar may take 24 hours to effectively sediment the larger particles (Song et al., 2019). The centrifugation process can be utilized to expedite this sedimentation process after mixing biochar into a solvent. Lian et al. (2020) used rice husk and straw to produce nano-biochar through the centrifugation method. The authors centrifuged the biomass at 10,000 rpm for 30 minutes to get nano-biochar with particle size of 100–350 nm. Plácido et al. (2019) used straw of crops (rice, sorghum) and manure obtained from dairy farms for nano-biochar production through centrifugation at the speed of 5,000 rpm for 20 minutes. Due to the cost of the membrane, membrane filtrations are speculated to be more expensive than sedimentation fractionation. The membrane filtration method can fulfill the requirement of specific particle nano-biochar (Dong et al., 2018).

9.3.3 CHARACTERIZATION

9.3.3.1 CHEMICAL CONFIGURATION OF BIOCHAR

Table 9.2 illustrates the results showing different physical and chemical properties of the biochars manufactured from rice straw (BC-R) and palm tree leaves (BC-P). The results demonstrate that the moisture percentage, saturation percentage (%SP), density, cation-exchange capacity (CEC), and ash percentage of type (BC-P) are higher than type (BC-R). CEC was determined in accordance with the proposed method known as the *Sumner and Miller method* (Sumner et al., 1996). BC-R showed higher N and P content than BC-P. Total K content was found to be greater in BC-P than in BC-R. The pH of BC-R and BC-P is demonstrated as 9.1 and 8.3, respectively. Pore surface area was measured by N_2 adsorption.

TABLE 9.2 Physical and Chemical Attributes of Biochar (BC-R and BC-P)

Attributes	BC-R	BC-P
SP	180	200
Moisture	4.75%	5.59
Total N	0.07%	0.04
Total P	0.07%	0.06
Total K	0.09%	0.32
density	0.13 g/cm^3	0.30
pH (1:5)	9.1	8.30
EC	3.55 dSm^{-1}	3.25
CEC	45 Cmol$_c$Kg^{-1}	51
Ash	55%	70

9.3.3.1.1 FT-IR OR FOURIER TRANSFORM INFRARED SPECTROSCOPY

For the identification of functional entities/groups present in the biochar types, FT-IR spectral investigation was done. This analysis was performed to identify functional groups that help in understanding the reactivity and efficiency of the biochar, which depends on the absorption/desorption of solutes present in the soil solution. Different peaks of functional groups present in BC-R were demonstrated according to FT-IR data. A peak of O-H alcohols and phenols (H-bonding) was observed at 3432.67 cm^{-1} (Peak No.1). C-H alkanes group was observed at positions 2923.56 cm^{-1} (Peak No.2), 1438.64 cm^{-1} (Peak No.4), and 708.712 cm^{-1} (Peak No.7). Peak No. 6 at 876.88 cm^{-1} represents C-H aromatics. The Carbonyl group (C=O) is demonstrated at 1620.88 cm^{-1} (Peak No. 3). Aliphatic amine was observed at 1104.05 cm^{-1} (Peak No. 5). Peak No.5 showed amines functional group (N-H) at 1625.7 cm^{-1} positions (Peak No. 5) and alkyl halides was observed at 65.181 cm^{-1} (Qian et al., 2013).

9.3.3.1.2 TEM OR TRANSMISSION ELECTRON MICROSCOPY

TEM helps in demonstrating the morphology of the particles and is also used for measuring the particle size. Spherical shapes of the particles of BC-R and BC-P were observed. The size of BC-R particles was observed between 5.59 nm and 11.3 nm, whereas BC-P particles were observed between 3.29 nm and 6.08 nm. Therefore, nano-biochar shows various advantages over macroscopic-size particles. These nanoparticles in biochar possess very high surface reactivity, infinite nanopores, and a very high adsorption capacity of ions.

9.3.3.1.3 NUTRIENTS SHOWING ADSORPTION/DESORPTION ON BIOCHAR

Adsorption isotherms illustrate the relationship between the quantities of nutrients (NH_4^+ and $H_2PO_4^-$) that are adsorbed on BC-R and BC-P and their concentrations at equilibrium. This relationship is observed in the form of straight lines, which are similar to a constant isotherm (Type C). Nutrients are adsorbed on the surface and within the existing pores present in the structure of biochar. It was observed from the adsorption isotherm that the absorption and removal capacity of the nutrient cation NH_4^+ and anion $H_2PO_4^-$ by BC-R is higher than BC-P, particularly at very high concentrations (equilibrium). The results indicate that with the addition of a very small amount of nano-biochar to the soil, the adsorption capacity of nutrients increases within the soil, which also helps in preserving nutrients against various losses.

9.3.3.1.4 NANO-COMPOSITES

The Freundlich isotherm model was studied for rice straw biochar (RSB), which can be used as a nano-absorbent material. The pores of RSB morphology were studied using TEM images. Large-sized pores were created using microporosity on the surface, which helped open the porous structure. It was observed that rice straw exhibits a disintegrated structure. The pyrolysis temperature was also increased from 600 to 800°C. The biochars produced at 800°C showed fragmented structures (Table 9.3) and were formed by cracking the parent biomass at very high temperatures (Khalil et al., 2019).

9.3.4 ELIMINATION OF HMS FROM TAINTED SOIL USING NANO-BIOCHAR COMPOSITES

Soil exhibiting different physical and chemical features can be affected by the accumulation of HMs in the soil (Wang et al., 2019). When nano-biochar nano-composites are added, the bioavailability of different HMs in the soil is affected. These nano-composites improve the pH and organic carbon levels

TABLE 9.3 Features of Rice Straw Biochar at 600 and 800°C

Item	600°C	800°C
Biochar yield	22.3%	15.3%
BET surface Area	63 $m^2.g^{-1}$	141 $m^2.g^{-1}$
Average pore size	6.5 nm	6.3 nm
Total pore volume	0.10 $cm^3.g^{-1}$	0.16 $cm^3.g^{-1}$
Bulk density	0.32 $gm.cm^{-3}$	0.44 $gm.cm^{-3}$
CEC	26.9	24.9
C	33.78%	20.17%
H	0.60%	0.25%
N	0.41%	0.25%
O	13.68%	3.71%
O/C	0.30	3.71
H/C	0.21	3.71
pH	10.04	10.84
Fe	170 mg Kg^{-1}	511.3 mg Kg^{-1}
Mn	112.9 mg Kg^{-1}	147.4 mg Kg^{-1}
Zn	17.7 mg Kg^{-1}	58.5 mg Kg^{-1}
Cu	41.5 mg Kg^{-1}	174.7 mg Kg^{-1}
As	39.8%	80.19%

of the soil. Heavy metals show different adsorptive properties when coming in contact with biochar-nanocomposites. A biochar/ALOOH is one of the nano-composites that was developed in collaboration with the Aluminum chloride biomass thermopyrolysis technique. After that, a nano-composite was developed by treating it with chitosan that contained amine functional groups and helped in the improvement of absorptivity of different toxic heavy metals, such as Pb (II), Cr (II), and Cd (II) on its surface. The removal of different heavy metals from the soil takes place using the following mechanism, which involves four processes.

9.3.4.1 ADSORPTION

During this process, large pore assemblies and surface zones are developed that allow effective adsorption of heavy metals (HMs) from the soil to the surface of nano-biochar. This adsorption property can be improved using nanobiochar-containing alkaline minerals and functional groups.

Chemisorption (chemical adsorption) takes place when different functional groups, such as carboxylic, phenolic, and hydroxylic groups, exchange their charged ions (-ve) with the positively charged impurities at the surface of nano-biochar (Lee et al., 2010). A temperature of 350–400C is needed to provide a good sorption capacity of nano-biochar.

9.3.4.2 COMPLEXATION

Complexation reactions are carried out using oxygen-bearing functional groups and carboxyl groups of biochar. Biochars containing hydroxyl groups and aromatic rings on their surface are responsible for carrying out complexation reactions with heavy metal ions.

9.3.4.3 PRECIPITATION

This process involves two pathways that help in removing HMs stress from the soil: when nano-biochar is added to the soil, soil pH value increases which helps in the precipitation of HMs from the contaminated soil. There is the reaction of inorganic components such as CO_3^{2-} and PO_4^{3-} in biochar with existing metal ions. This process is considered the main step in the removal of HMs by promoting the creation of solids on the nano-biochar surface during the process.

9.3.4.4 ION EXCHANGE

This method is required for the separation of toxic HMs from wastewater. The exchange of different metal cations in the soil is carried out using the high-cation exchange efficacy on the surface of nano-biochar. A brief summary of various studies involving the efficacy of nano-biochar or biochar nano-composites in the removal of heavy metals from soil has been represented in Table 9.4.

9.4 EFFICACY OF NANO-COMPOSITES AND NANO-BIOCHAR ON CROPS' LIFE CYCLE AND YIELD

9.4.1 NANO-COMPOSITES, NEED OF THE HOUR

Currently, the study on crop abiotic stress mostly focuses on an interdisciplinary approach involving both plant science and data science (Moulick et al., 2022b). The tools for studying biotic stress are available to identify genetic entities

TABLE 9.4 Efficiency in Removal of Heavy Metals by Nano-Biochar or Biochar Nano-Composites

Feedstock	Composite type	Heavy metals (Contaminants)	Removal/adsorption Capacity	References
Wood (pine tree or PB) natural hematite	HPB (Magnetic biochar-hematite-modified biochar	As(V)	HPB (429 mg Kg^{-1}) seems to be more efficient than PB (265 mg Kg^{-1}).	Wang et al. (2015)
Corn straw	BC	As	54.0–81.6% (phosphate-extractable).	Yu et al. (2017)
Sewage sludge		As	81.9% reduction noted in paddy soil (water-soluble).	Li et al. (2018)
Do	BC	As	Upgraded the plant biomass content (92.2%) and yield (175.1%) respectively. Minimize the As uptake in rice plant stem by 33.9%.	Khan et al. (2013)
Rice straw, Wheat straw and Kitchen waste		As	Efficacy in removing As content lies in the order of wheat straw was most efficient followed by rice straw and kitchen waste BC.	Kumar et al. (2021)
Straw of wheat	BC	As	83.7% efficacy in removing As content observed	Kumar, & Bhattacharya (2022)
Straw of corn	MBC (Manganese oxide-modified biochar)	As	14.36 mg g^{-1}	Yu et al. (2015)
Wood (Pine tree)	Manganese-oxide/biochar composite (Mn/BC)	Pb(II)	Removal efficiency of Pb (II) increased from 6.4 to 98.9%.	Wang et al. (2015)
Wood (Pine tree)	Manganese oxide–modified biochar (MBC)- $MnCl_2{\cdot}4H_2O$ (MPB), birnessite (BPB)	As(V) and Pb(II)	Variation noted in terms of efficacy.	Wang et al. (2015)
Eucalyptus leaf residue	Magnetic biochar	Pb(II)	Desorption efficiency found to be 84.1%	Wang et al. (2015)
Bamboo sawdust	BC	Pb	Immobilized 87% Pb (NH_4OAc-extractable)	Huang et al. (2018)

TABLE 9.4 *(Continued)*

Feedstock	Composite type	Heavy metals (Contaminants)	Removal/adsorption Capacity	References
Bamboo		Pb	Immobilized 73.2% Pb in sediment soil (HOAc-soluble)	Zhang et al. (2018)
Lead Chicken manure	BC	Pb	Significant enhancement in the rate of germination and root elongation by 360% and 189%, respectively. Followed by reduction in the bioavailability of Pb by 75.8%.	Ahmad et al. (2012)
Gaint miscanthus		Pb	Upgraded the plant nutrients availability status, and improved carbon sequestration. Improved the plant biomass content and minimize the bioavailability of Pb by 92%.	Houben et al. (2013)
Vegetable waste		Pb	Immobilized 87% Pb in agricultural soil (NH_4OAc-extractable)	Igalavithana et al. (2019)
Corn stover	Nano ZnS/ZnO modified biochar/zinc contaminated (Nanometal oxide)	Cr(VI), Cu(II) and Pb(II)	Effectivity in minimizing all the three pollutant content observed.	Li et al. (2018)
Rice straw	Ca-MBC (Calcium-based magnetic biochar)	As and Cd	Increased soil conductivity and pH on 32nd day.	Wu et al. (2020)
Red cedar (Thuja plicata) sawdust	Iron-modified magnetic biochar	As, Cd, and Pb	Effective in As, Cd, and Pb removal within 24 h. of application.	Wan et al. (2020)
Eichhornia crassipes	Chitosan modification of magnetic biochar (CMB)	Cr(VI)	Effective	Zhang et al. (2015)
Wheat straw	Wheat straw nano-biochar	Cd	Cd content lowered in treated soil group with comparison to untreated soil group ($P < 0.05$)	Liu et al. (2020)

TABLE 9.4 *(Continued)*

Feedstock	Composite type	Heavy metals (Contaminants)	Removal/adsorption Capacity	References
Walnut shell	Beta-cyclodextrin–chitosan modified walnut shell biochars (β-CCWB)	Cr(VI)	Exhibited removal efficiency about 93%.	Huang et al. (2016)
Sugarcane leafy trash	MgO hybrid carbonaceous composite	As(V), Pb(II) and MB	157 mg/g, 103 mg/g,	Li et al. (2018)
Red oak (RO), Switch grass (SG)	Iron-biochar (Magnetite biochar) (zero-valent iron (ZVI)- biochar complexes)	As	The order of effectivity as follows ZVI-RO (15.58 mg/g) trailing by ZVI-SG (7.92 mg/g)	Bakshi et al. (2018)
Rape straw	Alkaline treated biochar (BC–NaOH), $FeCl_3$ treated magnetic biochar (BC-FeOx), Biochar-$KMnO_4$ impregnation (BC-MnOx),	Cd(II)	BC-MnOx (81.10 mg/g) found highest in comparison to BC (32.74 mg/g) BC-MnOx > BC-NaOH > BC-FeOx	Li et al. (2017)
Rice husk	IPMN (Iron oxide permeated mesoporous rice-husk nano-biochars)	As	96% effectivity noted.	Nath et al. (2019)
Dendro wood	Dendro wood nano-biochar	Oxytetracycline (OTC)	16.9 and 113.2 mg/g removal observed.	Ramanayaka et al. (2020)

associated with agronomic traits of stress resilience in crops and surrounding factors affecting crop productivity. Apart from productivity, stress influence requires thorough monitoring throughout the crop life cycle. Records of wild crop domestication came from the initial stage of human civilization. But the current scenario for sustaining the 8 billion world population and eradicating hunger requires action on the 17 sustainable development goals (SDGs). As per a recent study by Lowery et al. (2019) maintaining production of such enormous feedstock exhaust 4×10^6 tons of pesticides, 2.7×10^9 m^3 of fresh-water, 187×10^6 tons of fertilizers, and more than 2×10^{15} British thermal units (BTUs) worth energy annually worldwide. All the magnitudes are for current scenario which is increasing with the population load.

9.4.2 APPLICATION OF NANO-COMPOSITES AND NANO-BIOCHAR

Agricultural nanoparticle products (agri-NPs) are used for HM mitigation in irrigation water, soil treatment, fertilizer and pesticide application. Nano-composite and nano-biochar have been attracting our attention. Even after crop harvest, agricultural waste treatment NPs have proven their worth. The effect of both natural NPs (NNP) and engineered NPs (ENP) on crops and the environment is currently under research (Ngˆo et al., 2014). In the current 17 SDGs taken up by the UN in 2015, NP application in nutrient management, soil and water management, hunger management, affordable and clean energy, and sustainable consumption patterns has a ubiquitous presence (Younis et al., 2021). Plenty of research is being conducted on the synthesis of nano-biochar to promote plant growth without contributing to climate change and alleviate water demands (Ighalo et al., 2020; Mukherjee et al., 2016; Mandal et al., 2021). Significant amounts of agri-NPs are used in integrated agro-ecosystem and animal husbandry practices (Peters et al., 2016). In terms of risk assessment, the framework for such agri-NP applications on agro-food production is limited to a laboratory environment. Meanwhile, long-term exposure studies on a field level, including the physico-chemical characteristics of ENPs on the human food system, are not yet completely known (Nurfatihah et al., 2019; Mishra et al., 2019; Makarenko et al., 2019).

The physico-chemical structure of any biochar determines the targeted areas for its application. For example, silicon-based biochar is effective in the cultivation of spinach under higher As exposure (Zama et al., 2018). Similarly, silver NPs have been proven to be effective in resisting potato ring rot disease-causing bacteria (Perfileva et al., 2016). Biochar provides mitigation in salt stress soil conditions, enhances nutrient availability, and

C sequestration in soil (Ding et al., 2016; Rodriguez et al., 2019). Magnetic Iron biochar provides greater yield in wheat production (Joseph et al., 2015). Similarly, phosphorus and sulfur ion incorporation elevate biochar activity and growth of lettuce (Gunes et al., 2014) and maize (Zhang et al., 2017), respectively. A rise in plant biomass, root length, and root mass by 32%, 52%, and 29%, respectively, is observed due to biochar application (Xiang et al., 2017). In Table 9.5, the major categories and sub-classifications of agri-NMs and the use of the items are briefly mentioned.

TABLE 9.5 Categories, Specific Product, and Use of NMs in Agricultural Practices

Sl. No	Categories of NMs	Products	Use
1.	Carbon	Carbon nanotubes (CNT)	Act as agrochemical deliverance system and mitigation of pesticide toxicity.
		Graphene	Microbial resistance.
		Fullerene	Antioxidant, high electron affinity.
		Nano-carbon fibers	
2.	Organic	Polymers	Micro-nutrients, encapsulation for nutrients.
3.	Metal and metal oxides	Au, Ag, Zn, Fe, Cu, Co NPs	Fertilizer, herbicides, fungicides, or bactericides.
		TiO_2, ZnO and other oxides	Fertilizer, use for water remediation, agrochemical coating, soil nutrient, antibiotic or pesticide.
		Nano-crystals	Anti-fungal, herbicide, anti-bacterial.
4.	Non-metals	Boron (B), nitrogen (N), phosphorus (P), and sulfur (S)	Nutrients for plant and soil microorganism, plant immunity enhancer.
5.	Hybrids	Organo-metal compounds, nano-composites	Nutrient deliverance, soil reclamation, sensor-based agri-practices.

Mg and Mn are both essential for salt stress tolerance (Guo et al., 2016). As Mg is ubiquitous in the chemical structures of enzymes and chlorophyll, it provides structural stability throughout the crop cycle under higher salinity exposure. Similarly, Mn has also been proven to improve nitrate reductase activity, ion homeostasis, and photosynthetic pigment formation in elevated salt levels (Rahman et al., 2016; Ye et al., 2020). Another study related to this by Ghassemi-Gole-zani et al. (2021) showed that biochar based on metal oxides of Mg and Mn mitigates Na uptake in plants under higher salinity in safflower and improves soil health as well (Sahoo et al., 2019; Moulick et al., 2020).

9.4.3 FATE OF NMS

Application of ENMs for higher crop yield also results in their incorporation into the ecosystem and soil for future cropping cycles. A study published on the fate of ENMs shows that 1.5% are in the atmosphere, 7% in water bodies and ecosystems, 28% in soil, and the highest (91%) in landfills (Keller et al., 2013). ENMs are a serious menace to the food chain due to the biomagnification factors in trophic transfer from soil to crop (Villena et al., 2018). As agriculture reports massive utilization of NMs during any crop cycle, the release of the same into the system has become another area under research. Phyto-nanotechnology studies the interaction between NMs and plant physiology. Encapsulations are designed to release agrochemicals or biomolecules at the target in specific dosages (Donia et al., 2019). Accumulation of metal oxide NMs entering the food chain in humans is undoubtedly a serious issue that needs to be addressed (Moulick et al., 2013; Arruda et al., 2015). Major studies on the fate of NMs are focused on the aquatic environment, but reports have shown significant deposition of ENMs in soil. NMs application in recycling, wastewater treatments, and agri NPs are majorly deposited in the soil system. Environmental conditions such as pH, electrolytes, and humic acid play an important role in the effectiveness of byproducts. Trophic transfer and ecotoxicological studies on NPs are currently crucial. Bioaccumulation factor (BAF), bioconcentration factor (BCF), biomagnification factor (BMF), and trophic transfer factor (TTF) are key parameters for predictive toxicity studies of an ENM. Hence, further use of these NMs requires safety measures along with efficiency studies.

9.5 AGRO-ENVIRONMENTAL ISSUES WITH NANO-COMPOSITES AND NANO-BIOCHAR

Soil contamination, which directly influences crop yield and productivity, is a prevalent and persistent problem in most parts of the globe. In developing countries, food security is promised through the extensive use of conventional chemical fertilizers, which results in HMs leaching (40–70%), leading to more significant economic losses. Over 5 million sites worldwide cover 1.2 billion ha of land, where HMs have contaminated the soil (Liu et al., 2018). As a result of HM contamination, not only crop yield but also qualitative aspects in the post-harvest phase are significantly influenced (Moulick et al., 2016a, 2017, 2020). Thus, biochar could be an effective management tool to address the current challenges in agriculture. Numerous scientific researchers have indicated that soil biochar applications enhance crop quality and quantity (Gujre et al., 2021; Mitra et al., 2016). However, compared to biochar, nano-biochar has distinctive and beneficial properties such as a greater specific surface area and hydrodynamic properties.

Furthermore, it is more hydrodynamically unstable and readily transported in the soil and aquatic ecosystem (Zhang et al., 2022). The metal mobility and toxicity in soil are primarily determined by metal speciation. The nano-biochar lowers the quantity of metal in exchangeable and carbonate fractions through greater reactivity, small particle size, and a large specific surface zone (Liu et al., 2020). The large surface area of the nano-biochar is chiefly due to the thermochemical degradation of alkyl and functional groups during the carbonization process (Tomczyk et al., 2020). Apart from that, nano-biochar has core-shell structures that may improve incapacitation via HM reduction. The oxide shell adsorbs heavy metals on its surface through electrostatic actions and surface complexation, while the metallic core acts as an electron source for removal (Liu et al., 2014).

A study by Tyc et al. (2017) indicated that the pedosphere biogeochemical behavior of nano-biochar and heavy metals is governed by bioactive compounds such as lipopeptides, siderophores, and exopolysaccharides. Certain studies have suggested that nano-biochar produced at 350°C from wheat straw could potentially act as a slow fertilizer for crops, while also reducing leaching processes (Khan et al., 2021). These intrinsic features of nano-biochar help improve crop health and provide high nutrient content over an extended period of time. A study on nano-biochar revealed that as it ages, raw carbon nanoparticles (rCNPs) slowly oxidize in the atmosphere, leading to an increase in their porous surface structure. This porous surface structure enhances the adsorptive capacity of the rCNPs and promotes wheat growth (Saxena et al., 2014). Nano-biochar prepared at higher temperatures also effectively mitigates the toxicity of Cd compared to nano-biochar prepared at lower temperatures. Additionally, it improves root vitality, chlorophyll content, and antioxidative enzyme activities (Khan et al., 2021). Globally, the agriculture sector experiences significant crop yield reductions annually due to insect pests and diseases, ranging from 20 to 40%, which exacerbates food insecurity. However, nano-biochar produced from rice straw could effectively reduce pest-induced disease incidence and severity by up to 50% and 22.5%, respectively, caused by *Xanthomonas campestris PV. Vesicatoria* (Aftab et al., 2022). Rangaraj et al. (2014) noted that the process utilized silica nanoparticles as an effective pest control tool against pests like *Fusarium oxysporum* and *Aspergillus niger* in maize. Corn straw nano-biochar formed at a lower temperature (350°C) had an encouraging effect on the plant as it increased the reactive oxygen species (ROS) up to 20.3%, triggering the ethylene pathway and thereby increasing the active immunity of *N. benthamiana* from *P. nicotianae* infection (Kong et al., 2022). For efficient P absorption, salinity played a critical role, and the application of nano-biochar (0.20%) increased the Freundlich constant (KF) from an initial value of 64.60 mg/kg to 497.60 mg/kg in the soil (Mahmoud et al., 2022). Furthermore, Lateef et al. (2019) reported that corn-derived nano-composite biochar fixed the leaching issue after applying the traditional fertilizer. An actual rice paddy column application of nano-cerium oxide functionalized maize straw biochar (Ce-MSB) was tested. The field experiment revealed that a higher material ratio of Ce-MSB enhances the phosphate adsorption capacity of plants up to 78 mg/g (Feng et al., 2017). The application of nano-biochar is vast and could be used as a nano-fertilizer. Particularly, research suggested that applying nano-biochar increases the height of the tested plants, viz., *Ocimum basilicum*, by up to 1.5% (Zahedifar & Najafian, 2017). Furthermore, recent research investigated that the application of the areca nut husk biochar of 15 t ha^{-1} improves the chlorophyll content, biomass, and surface area of the vetiver leaves (Gujre et al., 2022). The application of nano-biochar not only influences plant morphology and physiology but also affects the soil's physical properties. For instance, in a dose-responsive manner, the application of nano-biochar decreased the cumulative infiltration value from 3.75% to 13.75% (Chen et al., 2022). Several researchers worldwide have portrayed that nano-biochar is superior in the adsorption of agrochemicals and toxic heavy metals compared to bulk biochar applications (Kalia et al., 2011). It is a fact that biochar contains ample amounts of hydrophilic micro and nano-carbons fitted with oxo-, a functional group that could be efficiently taken up by the plant. Such intrinsic properties assist in the growth of the plant (Bhati et al., 2018). Nano-biochar is considered a sustainable strategy that could effectively address the issues of food security and greenhouse gas emissions. The developed material could be applied with different phosphate-solubilizing bacteria and agrochemicals for sustainable farming. However, further studies need to be conducted to decipher the dynamics of nano-biochar in the agroecosystem.

9.6 CONCLUSION

Nano-composites and nano-biochar have a promising future in the field of sustainable agriculture due to their potential role in remediating HMs-contaminated soil and promoting good plant health and crop yield even under considerable levels of heavy metal stress. The preparation of nano-composites and nano-biochar can be done by simple methods using various biomass sources, which could be a feasible option for agricultural waste recycling. Burning agricultural residues often poses severe environmental risks. However, bulk production of nano-composites and nano-biochar for application in farmers' fields could be a challenge, which can be resolved with the application of advanced technologies. Policy support is necessary to make these technologies affordable to farmers, especially in developing countries. Sufficient infrastructure is also needed for the collection and transport of biomass from fields to the production facility. Although the beneficial effects of nano-composites and nano-biochar are confirmed by several studies, only partial information is available on the long-term consequences of these materials on crop life cycles and the sustainability of agroecosystems and biodiversity. Detailed insights into the molecular mechanisms involving nano-composites and nano-biochar in the responses of the plants are yet to be acquired.

KEYWORDS

- **crop response**
- **heavy metal contamination**
- **nano-biochar**
- **nano-composite**
- **sustainable agriculture**

REFERENCES

Aftab, Z. E. H., Aslam, W., Aftab, A., Shah, A. N., Akhter, A., Fakhar, U., Siddiqui, I., et al., (2022). Incorporation of engineered nanoparticles of biochar and fly ash against bacterial leaf spot of pepper. *Scientific Reports, 12*(1), 1–13.

Ahmad, M., Lee, S. S., Yang, J. E., Ro, H. M., Lee, Y. H., & Ok, Y. S., (2012). Effects of soil dilution and amendments (mussel shell, cow bone, and biochar) on Pb availability and phytotoxicity in military shooting range soil. *Ecotoxicology and Environmental Safety, 79*, 225–231.

Ahmad, M., Ok, Y. S., Kim, B. Y., Ahn, J. H., Lee, Y. H., Zhang, M., Moon, D. H., et al., (2016). Impact of soybean stover-and pine needle-derived biochars on Pb and As mobility, microbial community, and carbon stability in a contaminated agricultural soil. *Journal of Environmental Management, 166*, 131–139.

Ahmad, M., Rajapaksha, A. U., Lim, J. E., Zhang, M., Bolan, N., Mohan, D., Vithanage, M., et al., (2014). Biochar as a sorbent for contaminant management in soil and water: A review. *Chemosphere, 99*, 9–33.

Alexander, P., Brown, C., Arneth, A., Finnigan, J., Moran, D., & Rounsevell, M. D. A., (2017). Losses, inefficiencies and waste in the global food system. *Agricultural Systems, 153*, 190–200.

Ali, H., Khan, E., & Ilahi, I., (2019). Environmental chemistry and ecotoxicology of hazardous heavy metals: Environmental persistence, toxicity, and bioaccumulation. *Journal of Chemistry, 2019.*

Alia, N., Sardar, K., Said, M., Salma, K., Sadia, A., Sadaf, S., & Toqeer, A., (2015). Toxicity and bioaccumulation of heavy metals in spinach (*Spinacia oleracea*) grown in a controlled environment. *International Journal of Environmental Research and Public Health, 12*(7), 7400–7416.

Amer, M., & Elwardany, A., (2020). Biomass carbonization. In: Al Qubeissi, M., El-Kharouf, A., & Soyhan, H. S., (eds.), *Renewable Energy – Resources, Challenges and Applications* (pp. 1–22). Rijeka: IntechOpen.

Anupama, & Khare, P., (2021). A comprehensive evaluation of inherent properties and applications of nano-biochar prepared from different methods and feedstocks. *Journal of Cleaner Production, 320*, 128759.

Arruda, S. C. C., Silva, A. L. D., Galazzi, R. M., Azevedo, R. A., & Arruda, M. A. Z., (2015). Nanoparticles applied to plant science: A review. *Talanta, 131*, 693–705.

Awasthi, J. P., Saha, B., Chowardhara, B., Borgohain, P., Sahoo, S., Tanti, B., & Panda, S. K., (2022). Differential response to acidic pH in rice seedlings. *SAINS TANAH-Journal of Soil Science and Agroclimatology, 19*(1), 12–18.

Baker, S., Volova, T., Prudnikova, S. V., Satish, S., & Prasad, M. N. N., (2017). Nanoagroparticles emerging trends and future prospect in modern agriculture system. *Environ. Toxicol. Pharmacol., 53,* 10–17. https://doi.org/10.1016/j.etap.2017.04.012.

Bakshi, S., Banik, C., Rathke, S. J., & Laird, D. A., (2018). Arsenic sorption on zero-valent iron-biochar complexes. *Water Research, 137*, 153–163.

Beesley, L., Moreno-Jim´enez, E., Gomez-Eyles, J. L., Harris, E., Robinson, B., & Sizmur, T., (2011). A review of biochars potential role in the remediation, revegetation and restoration of contaminated soils. *Environmental Pollution, 159*(12), 3269–3282. https://doi.org/10.1016/j.envpol.2011.07.023.

Bhati, A., Tripathi, K. M., Singh, A., Sarkar, S., & Sonkar, S. K., (2018). Exploration of nano carbons in relevance to plant systems. *New Journal of Chemistry, 42*(20), 16411–16427.

Bhunia, P., (2017). Environmental toxicants and hazardous contaminants: Recent advances in technologies for sustainable development. *Journal of Hazardous, Toxic, and Radioactive Waste, 21*(4), 02017001.

Bridgwater, A. V., (2007). The production of biofuels and renewable chemicals by fast pyrolysis of biomass. *International Journal of Global Energy Issues, 27*(2), 160–203.

Bridgwater, A. V., Toft, A. J., & Brammer, J. G., (2002). A techno-economic comparison of power production by biomass fast pyrolysis with gasification and combustion. *Renewable and Sustainable Energy Reviews, 6*(3), 181–246.

Chen, M., Alim, N., Zhang, Y., Xu, N., & Cao, X., (2018). Contrasting effects of biochar nanoparticles on the retention and transport of phosphorus in acidic and alkaline soils. *Environmental Pollution, 239*, 562–570.

Chen, M., Wang, D., Yang, F., Xu, X., Xu, N., & Cao, X., (2017). Transport and retention of biochar nanoparticles in a paddy soil under environmentally-relevant solution chemistry conditions. *Environmental Pollution, 230*, 540–549.

Chen, X., Duan, M., Zhou, B., & Cui, L., (2022). Effects of biochar nanoparticles as a soil amendment on the structure and hydraulic characteristics of a sandy loam soil. *Soil Use and Management, 38*(1), 836–849.

Chen, Y. X., He, Y. F., Luo, Y. M., Yu, Y. L., Lin, Q., & Wong, M. H., (2003). Physiological mechanism of plant roots exposed to cadmium. *Chemosphere, 50*(6), 789–793.

Choudhury, S., & Moulick, D., (2022). *Response of Field Crops to Abiotic Stress: Current Status and Future Prospects.* CRC Press. ISBN: 9781032191966.

Choudhury, S., Mazumder, M. K., Moulick, D., Sharma, P., Tata, S. K., Ghosh, D., Ali, M. H., et al., (2022a). A Computational study of the role of secondary metabolites for mitigation of acid soil stress in cereals using dehydroascorbate and mono – dehydroascorbate reductases. *Antioxidant, 11*(3), 458; https://doi.org/10.3390/antiox11030458.

Choudhury, S., Moulick, D., & Mazumder, M. K., (2021a). Secondary metabolites protect against metal and metalloid stress in rice: An in-silico investigation using dehydroascorbate reductase. *Acta Physiologiae Plantarum., 43*(1), 1–10. https://doi.org/10.1007/s1173-020-03173-2.

Choudhury, S., Moulick, D., Mazumder, M. K., Pattnaik, B. K., Ghosh, D., Vemireddy, L. R., Aldhahrani, A., et al., (2022b). An *in vitro* and in silico perspective study of seed priming with zinc on the phytotoxicity and accumulation pattern of arsenic in rice seedlings. *Antioxidant Ntioxidants, 11*(8), 1500. https://doi.org/10.3390/antiox11081500.

Choudhury, S., Sharma, P., Moulick, D., & Mazumder, M. K., (2021b). Unrevealing metabolomics for abiotic stress adaptation and tolerance in plants. *Journal of Crop Science and Biotechnology, 24,* 479–493. https://doi.org/10.1007/s12892-021-00102-8.

Chowardhara, B., Borgohain, P., Saha, B., Awasthi, J. P., & Panda, S. K., (2020). Differential oxidative stress responses in *Brassica juncea* (L.) Czern and Coss cultivars induced by cadmium at germination and early seedling stage. *Acta Physiologiae Plantarum, 42*(7), 1–12.

Chowardhara, B., Borgohain, P., Saha, B., Awasthi, J. P., Moulick, D., & Panda, S. K., (2019a). Assessment of phytotoxicity of zinc on Indian mustard (*Brassica juncea*) varieties during germination and early seedling growth. *Annals of Plant and Soil Research, 21*(3), 239–244.

Chowardhara, B., Borgohain, P., Saha, B., Awasthi, J. P., Moulick, D., & Panda, S. K., (2019b). Phytotoxicity of Cd and Zn on three popular Indian mustard varieties during germination and early seedling growth. *Biocatalysis and Agricultural Biotechnology, 21*, 101349.

Ciscato, M., Vangronsveld, J., & Valcke, R., (1999). Effects of heavy metals on the fast chlorophyll fluorescence induction kinetics of photosystem II: A comparative study. *Zeitschrift für Naturforschung C, 54*(9, 10), 735–739.

Demirbas, A., (2004). Effects of temperature and particle size on bio-char yield from pyrolysis of agricultural residues. *Journal of Analytical and Applied Pyrolysis, 72*(2), 243–248.

Development Initiatives, (2017). *Global Nutrition Report 2017: Nourishing the SDGs.* Development Initiatives, Bristol. https://doi.org/10.1038/s41893-019-0248-8.

Ding, Y., Liu, Y., Liu, S., Li, Z., Tan, X., Huang, X., Zeng, G., Zhou, L., & Zheng, B., (2016). Biochar to improve soil fertility. A review. *Agron. Sustain. Dev., 36*, 1–18. https:// doi.org/10.1007/s13593-016-0372-z.

Dong, X., He, L., Liu, Y., & Piao, Y., (2018). Preparation of highly conductive biochar nanoparticles for rapid and sensitive detection of 17β-estradiol in water. *Electrochimica Acta, 292*, 55–62.

FAO, (2019). *The State of Food Security and Nutrition in the World: Safeguarding Against Economic Slowdowns and Downturns*. Food and Agriculture Organizations of the United Nations, Rome.

FAO, (2022). *How to Feed the World in 2050?* https://www.fao.org/fileadmin/templates/wsfs/docs/expert_paper/How_to_Feed_the_World_in_2050.pdf (accessed on 25 April 2024).

Feng, Y., Lu, H., Liu, Y., Xue, L., Dionysiou, D. D., Yang, L., & Xing, B., (2017). Nano-cerium oxide functionalized biochar for phosphate retention: Preparation, optimization and rice paddy application. *Chemosphere, 185*, 816–825.

Foley, J. A., Ramankutty, N., Brauman, K. A., Cassidy, E. S., Gerber, J. S., Johnston, M., Mueller, N. D., et al., (2011). Solutions for a cultivated planet. *Nature, 478*, 337–342. https://doi.org/10.1038/nature10452.

Funke, A., & Ziegler, F., (2010). Hydrothermal carbonization of biomass: A summary and discussion of chemical mechanisms for process engineering. *Biofuels, Bioproducts and Biorefining, 4*(2), 160–177.

Gámiz, B., Cox, L., Hermosín, M. C., Spokas, K., & Celis, R., (2017). Assessing the effect of organoclays and biochar on the fate of abscisic acid in soil. *Journal of Agricultural and Food Chemistry, 65*(1), 29–38.

Ghassemi-Golezani, K., Farhangi-Abriz, S., & Abdoli, S., (2021). How can biochar-based metal oxide nanocomposites counter salt toxicity in plants? *Environ. Geochem. Health, 43*, 2007–2023. https://doi.org/10.1007/s10653-020-00780-3.

Ghosh, D., Brahmachari, K., Brestic, M., Ondrisik, P., Hossain, A., Skalicky, M., Sarkar, S., et al., 2020a. Integrated weed and nutrient management improve yield, nutrient uptake and economics of maize in the rice-maize cropping system of Eastern India. *Agronomy, 10*(12), 1906. https://doi.org/10.3390/agronomy 10121906.

Ghosh, D., Brahmachari, K., Das, A., Hassan, M. M., Mukherjee, P. K., Sarkar, S., Dinda, N. K., et al., (2021). Assessment of energy budgeting and its indicator for sustainable nutrient and weed management in a rice-maize-green gram cropping system. *Agronomy, 11*(1), 166. https://doi.org/10.3390/agronomy11010166.

Ghosh, D., Brahmachari, K., Skalicky, M., Hossain, A., Sarkar, S., Dinda, N. K., Das, A., et al., 2020b. Nutrients supplementation through organic manures influence the growth of weeds and maize productivity. *Molecules, 25*(21), 4924. https://doi.org/10.3390/molecules 25214924.

Ghosh, D., Brahmachari, K., Skalický, M., Roy, D., Das, A., Sarkar, S., Moulick, D., et al., (2022). The combination of organic and inorganic fertilizers influence the weed growth, productivity and soil fertility of monsoon rice. *PLoS One, 17*(1), e0262586. https://doi.org/10.1371/journal.pone.0262586.

Gujre, N., Mitra, S., Agnihotri, R., Sharma, M. P., & Gupta, D., (2022). Novel agrotechnological intervention for soil amendment through areca nut husk biochar in conjunction with vetiver grass. *Chemosphere, 287*, 132443.

Gujre, N., Soni, A., Rangan, L., Tsang, D. C., & Mitra, S., (2021). Sustainable improvement of soil health utilizing biochar and arbuscular mycorrhizal fungi: A review. *Environmental Pollution, 268*, 115549.

Gunes, A., Inal, A., Taskin, M. B., Sahin, O., Kaya, E. C., & Atakol, A., (2014). Effect of phosphorus-enriched biochar and poultry manure on growth and mineral composition of lettuce (*Lactuca sativa* L. cv.) grown in alkaline soil. *Soil Use Manag., 30*, 182–188. https://doi.org/10.1111/sum.12114.

Guo, W., Nazim, H., Liang, Z., & Yang, D., (2016). Magnesium deficiency in plants: An urgent problem. *Crop J., 4*, 83–91. https://doi.org/10.1016/j.cj.2015.11.003.

Gustavsson, J., Cederberg, C., Sonesson, U., Van, O. R., & Meybeck, A., (2011). *Global Food Losses and Food Waste – Extent, Causes and Prevention.* FAO, Rome.

Hazra, S., Swain, D. K., & Bhadoria, P. B. S., (2019). Wheat grown under elevated CO_2 was more responsive to nitrogen fertilizer in Eastern India. *European Journal of Agronomy, 105*, 1–12.

Hossain, M. K., Strezov, V., Chan, K. Y., Ziolkowski, A., & Nelson, P. F., (2011). Influence of pyrolysis temperature on production and nutrient properties of wastewater sludge biochar. *Journal of Environmental Management, 92*(1), 223–228.

Houben, D., Evrard, L., & Sonnet, P., (2013). Beneficial effects of biochar application to contaminated soils on the bioavailability of Cd, Pb and Zn and the biomass production of rapeseed (*Brassica napus* L.). *Biomass and Bioenergy, 57*, 196–204.

Huang, D., Deng, R., Wan, J., Zeng, G., Xue, W., Wen, X., Zhou, C., et al., (2018). Remediation of lead-contaminated sediment by biochar-supported nano-chlorapatite: Accompanied with the change of available phosphorus and organic matters. *Journal of Hazardous Materials, 348*, pp.109–116.

Huang, Q., Song, S., Chen, Z., Hu, B. W., Chen, J. R., & Wang, X. K., (2019). Biochar-based materials and their applications in removal of organic contaminants from wastewater: State-of-the-art review. *Biochar, 1*, 45–73.

Huang, X., Liu, Y., Liu, S., Tan, X., Ding, Y., Zeng, G., Zhou, Y., et al., (2016). Effective removal of Cr (VI) using β-cyclodextrin–chitosan modified biochars with adsorption/reduction bifuctional roles. *RSC Advances, 6*(1), 94–104.

Igalavithana, A. D., Kwon, E. E., Vithanage, M., Rinklebe, J., Moon, D. H., Meers, E., Tsang, D. C., & Ok, Y. S., (2019). Soil lead immobilization by biochars in short-term laboratory incubation studies. *Environment International, 127*, 190–198.

Ighalo, J. O., (2020). Biomass to biochar conversion for agricultural and environmental applications in Nigeria: Challenges, peculiarities and prospects. *Met. Mater. Int., 2*, 111–116. https://doi.org/10.33263/Materials22.111116.

Jadhav, S. S., Kumari, R., Mahtha, S. K., Purama, R. K., Lamba, V., & Yadav, G., (2022). Metabolomics and molecular physiology perspective for drought and salinity stress tolerance. In: *Response of Field Crops to Abiotic Stress* (pp. 153–166). CRC Press.

Jensen, P. A., Frandsen, F. J., Dam-Johansen, K., & Sander, B., (2000). Experimental investigation of the transformation and release to gas phase of potassium and chlorine during straw pyrolysis. *Energy & Fuels, 14*(6), 1280–1285.

Jiang, C., Bo, J., Xiao, X., Zhang, S., Wang, Z., Yan, G., Wu, Y., Wong, C., & He, H., (2020). Converting waste lignin into nano-biochar as a renewable substitute of carbon black for reinforcing styrene-butadiene rubber. *Waste Management, 102*, 732–742.

John, R., Ahmad, P., Gadgil, K., & Sharma, S., (2009). Heavy metal toxicity: Effect on plant growth, biochemical parameters, and metal accumulation by *Brassica juncea* L. *International Journal of Plant Production, 3*(3), 66–75. SID. https://sid.ir/paper/314576/en (accessed on 6 May 2024).

Joseph, S., Anawar, H. M., Storer, P., Blackwell, P., Chee, C. H. I. A., Yun, L. I. N., Munroe, P., et al., (2015). Effects of enriched biochars containing magnetic iron nanoparticles on mycorrhizal colonization, plant growth, nutrient uptake and soil quality improvement. *Pedosphere, 25*, 749–760. https://doi.org/10.1016/S1002-0160(15)30056-4.

Kalia, S., Dufresne, A., Cherian, B. M., Kaith, B. S., Avérous, L., Njuguna, J., & Nassiopoulos, E., (2011). Cellulose-based bio-and nanocomposites: A review. *International Journal of Polymer Science, 2011.*

Khan, H. A., Naqvi, S. R., Mehran, M. T., Khoja, A. H., Niazi, M. B. K., Juchelková, D., & Atabani, A., (2021). A performance evaluation study of nano-biochar as a potential slow-release nano-fertilizer from wheat straw residue for sustainable agriculture. *Chemosphere, 285*, 131382.

Khan, S., Chao, C., Waqas, M., Arp, H. P. H., & Zhu, Y. G., (2013). Sewage sludge biochar influence upon rice (*Oryza sativa* L) yield, metal bioaccumulation and greenhouse gas emissions from acidic paddy soil. *Environmental Science & Technology, 47*(15), 8624–8632.

Khorram, M. S., Zhang, Q., Lin, D., Zheng, Y., Fang, H., & Yu, Y., (2016). Biochar: A review of its impact on pesticide behavior in soil environments and its potential applications. *Journal of Environmental Sciences, 44*, 269–279.

Kim, H. B., Kim, S. H., Jeon, E. K., Kim, D. H., Tsang, D. C., Alessi, D. S., Kwon, E. E., & Baek, K., (2018). Effect of dissolved organic carbon from sludge, rice straw and spent coffee ground biochar on the mobility of arsenic in soil. *Science of the Total Environment, 636*, 1241–1248.

Kong, M., Liang, J., White, J. C., Elmer, W. H., Wang, Y., Xu, H., He, W., et al., (2022). Biochar nanoparticle-induced plant immunity and its application with the elicitor methoxyindole in *Nicotiana benthamiana. Environmental Science: Nano, 9*(9), 3514–3524.

Koo, B. J., Chang, A. C., Crowley, D. E., Page, A. L., & Taylor, A., (2013). Availability and plant uptake of biosolid-borne metals. *Applied and Environmental Soil Science, 2013.*

Kookana, R. S., Sarmah, A. K., Van, Z. L., Krull, E., & Singh, B., (2011). Biochar application to soil: Agronomic and environmental benefits and unintended consequences. In: Sparks, D. L., (ed.), *Advances in Agronomy* (pp. 103–143). Academic Press, Elsevier. https://doi.org/10.1016/B978-0-12-385538-1.00003-2.

Kumar, A., & Bhattacharya, T., (2022). Removal of arsenic by wheat straw biochar from soil. *Bulletin of Environmental Contamination and Toxicology, 108*(3), 415–422.

Kumar, A., Bhattacharya, T., Shaikh, W. A., Roy, A., Mukherjee, S., & Kumar, M., (2021). Performance evaluation of crop residue and kitchen waste-derived biochar for eco-efficient removal of arsenic from soils of the Indo-Gangetic plain: A step towards sustainable pollution management. *Environmental Research, 200*, 111758.

Kumar, M., Xiong, X., Wan, Z., Sun, Y., Tsang, D. C. W., Gupta, J., Gao, B., et al., (2020). Ball milling as a mechanochemical technology for fabrication of novel biochar nanomaterials. *Bioresource Technology, 312*, 123613.

Lahori, A. H., Zhanyu, G. U. O., Zhang, Z., Ronghua, L. I., Mahar, A., Awasthi, M. K., & Jiang, S., (2017). Use of biochar as an amendment for remediation of heavy metal-contaminated soils: Prospects and challenges. *Pedosphere, 27*(6), 991–1014.

Law, P. N. K., Rahman, M. R., Hamdan, S., Jayamani, E., Bakri, M. K. B., & Sanaullah, K., (2020). Synthesis and characterization of micro-nano carbon filler from jatropha seeds. *BioResources, 15*(2), 3237–3251.

Lee, J. W., Kidder, M., Evans, B. R., Paik, S., Buchanan, I. A. C., Garten, C. T., & Brown, R. C., (2010). Characterization of biochars produced from cornstovers for soil amendment. *Environmental Science & Technology, 44*(20), 7970–7974.

Li, B., Yang, L., Wang, C. Q., Zhang, Q. P., Liu, Q. C., Li, Y. D., & Xiao, R., (2017). Adsorption of Cd (II) from aqueous solutions by rape straw biochar derived from different modification processes. *Chemosphere, 175*, 332–340.

Li, C., Zhang, L., Gao, Y., & Li, A., (2018). Facile synthesis of nano ZnO/ZnS modified biochar by directly pyrolyzing of zinc contaminated corn stover for Pb (II), Cu (II) and Cr (VI) removals. *Waste Management, 79*, 625–637.

Li, G., Khan, S., Ibrahim, M., Sun, T. R., Tang, J. F., Cotner, J. B., & Xu, Y. Y., (2018). Biochars induced modification of dissolved organic matter (DOM) in soil and its impact on mobility and bioaccumulation of arsenic and cadmium. *Journal of Hazardous Materials, 348*, 100–108.

Li, L., Zhang, K., Chen, L., Huang, Z., Liu, G., Li, M., & Wen, Y., (2017). Mass preparation of micro/nano-powders of biochar with water-dispersibility and their potential application. *New Journal of Chemistry, 41*(18), 9649–9657.

Li, R., Liang, W., Wang, J. J., Gaston, L. A., Huang, D., Huang, H., Lei, S., et al., (2018). Facilitative capture of As (V), Pb (II) and methylene blue from aqueous solutions with MgO hybrid sponge-like carbonaceous composite derived from sugarcane leafy trash. *Journal of Environmental Management, 212*, 77–87.

Li, R., Zhang, Y., Deng, H., Zhang, Z., Wang, J. J., Shaheen, S. M., Xiao, R., et al., (2020). Removing tetracycline and Hg(II) with ball-milled magnetic nanobiochar and its potential on polluted irrigation water reclamation. *Journal of Hazardous Materials, 384*, 121095.

Li, W., Zhu, X., He, Y., Xing, B., Xu, J., & Brookes, P. C., (2013). Enhancement of water solubility and mobility of phenanthrene by natural soil nanoparticles. *Environmental Pollution, 176*, 228–233.

Lian, F., & Xing, B., (2017). Black carbon (biochar) in water/soil environments: Molecular structure, sorption, stability, and potential risk. *Environmental Science & Technology, 51*(23), 13517–13532.

Lian, F., Yu, W., Zhou, Q., Gu, S., Wang, Z., & Xing, B., (2020). Size matters: Nano-biochar triggers decomposition and transformation inhibition of antibiotic resistance genes in aqueous environments. *Environmental Science & Technology, 54*(14), 8821–8829.

Libra, J. A., Ro, K. S., Kammann, C., Funke, A., Berge, N. D., Neubauer, Y., Titirici, M. M., et al., (2011). Hydrothermal carbonization of biomass residuals: A comparative review of the chemistry, processes and applications of wet and dry pyrolysis. *Biofuels, 2*(1), 71–106.

Lin, R., Wang, X., Luo, Y., Du, W., Guo, H., & Yin, D., (2007). Effects of soil cadmium on growth, oxidative stress and antioxidant system in wheat seedlings (*Triticum aestivum* L.). *Chemosphere, 69*(1), 89–98.

Lin, Y., Ma, X., Peng, X., Yu, Z., Fang, S., Lin, Y., & Fan, Y., (2016). Combustion, pyrolysis and char CO_2-gasification characteristics of hydrothermal carbonization solid fuel from municipal solid wastes. *Fuel, 181*, 905–915.

Liu, F., Yang, J., Zuo, J., Ma, D., Gan, L., Xie, B., Wang, P., & Yang, B., (2014). Graphene-supported nanoscale zero-valent iron: Removal of phosphorus from aqueous solution and mechanistic study. *Journal of Environmental Sciences, 26*(8), 1751–1762.

Liu, G., Zheng, H., Jiang, Z., Zhao, J., Wang, Z., Pan, B., & Xing, B., (2018). Formation and physicochemical characteristics of nano biochar: Insight into chemical and colloidal stability. *Environmental Science & Technology, 52*(18), 10369–10379.

Liu, J., Jiang, J., Meng, Y., Aihemaiti, A., Xu, Y., Xiang, H., Gao, Y., & Chen, X., (2020). Preparation, environmental application and prospect of biochar-supported metal nanoparticles: A review. *Journal of Hazardous Materials, 388*, 122026.

Liu, L., Li, W., Song, W., & Guo, M., (2018). Remediation techniques for heavy metal-contaminated soils: Principles and applicability. *Science of the Total Environment, 633*, 206–219.

Liu, W. J., Jiang, H., & Yu, H. Q., (2015). Development of biochar-based functional materials: Toward a sustainable platform carbon material. *Chemical Reviews, 115*(22), 12251–12285.

Liu, W., Li, Y., Feng, Y., Qiao, J., Zhao, H., Xie, J., Fang, Y., Shen, S., & Liang, S., (2020). The effectiveness of nanobiochar for reducing phytotoxicity and improving soil remediation in cadmium-contaminated soil. *Scientific Reports, 10*(1), 1–10.

Liu, X., Wu, J., & Xu, J., (2006). Characterizing the risk assessment of heavy metals and sampling uncertainty analysis in paddy field by geostatistics and GIS. *Environmental Pollution, 141*(2), pp.257–264.

Lonappan, L., Rouissi, T., Das, R. K., Brar, S. K., Ramirez, A. A., Verma, M., Surampalli, R. Y., & Valero, J. R., (2016). Adsorption of methylene blue on biochar microparticles derived from different waste materials. *Waste Management, 49*, 537–544.

Lowry, G. V., Avellan, A., & Gilbertson, L. M., (2019). Opportunities and challenges for nanotechnology in the agri-tech revolution. *Nat. Nanotechnol., 14*, 517–522. https://doi.org/10.1038/s41565-019-0461-7.

Lyu, H., Gao, B., He, F., Zimmerman, A. R., Ding, C., Huang, H., & Tang, J., (2018). Effects of ball milling on the physicochemical and sorptive properties of biochar: Experimental observations and governing mechanisms. *Environmental Pollution, 233*, 54–63.

Mahmood, S., Hussain, A., Saeed, Z., & Athar, M., (2005). Germination and seedling growth of corn (*Zea mays* L.) under varying levels of copper and zinc. *International Journal of Environmental Science & Technology, 2*(3), 269–274.

Mahmoud, E., El Baroudy, A., Ali, N., & Sleem, M., (2020). Spectroscopic studies on the phosphorus adsorption in salt-affected soils with or without nano-biochar additions. *Environmental Research, 184*, 109277.

Maitra, S., Mondal, T., Hossain, A., Kalasare, R. S., Praharaj, S., Joy, J. M. M., Shankar, T., et al., (2022). Impact of climate change on growth and productivity of major field crops. In: *Response of Field Crops to Abiotic Stress* (pp. 217–226). CRC Press.

Makarenko, N. A., & Makarenko, V. V., (2019). Nanotechnologies in crop cultivation: Ecotoxicological aspects. *Biosyst. Divers., 27*, 148–155. https://doi.org/ 10.15421/011920.

Makshut, N. A., Ngaini, Z., Wahi, R., Hussain, H., Mahmut, N. I., & Bahrin, N. Q., (2020). Nano-sized adsorbent from pyrolyzed sago activated sludge for removal of Pb (II) from aqueous solution. *Pertanika Journal of Science & Technology, 28*(3).

Mandal, S., Pu, S., Adhikari, S., Ma, H., Kim, D. H., Bai, Y., et al., (2021). Progress and future prospects in biochar composites: Application and reflection in the soil environment. *Crit. Rev. Environ. Sci. Technol., 51*, 219–271. https://doi.org/ 10.1080/10643389.2020.1713030.

Manzoor, J., Sharma, M., & Wani, K. A., (2018). Heavy metals in vegetables and their impact on the nutrient quality of vegetables: A review. *Journal of Plant Nutrition, 41*(13), 1744–1763.

Mazumder, M. K., Moulick, D., & Choudhury, S., (2022). Iron (Fe^{3+})-mediated redox responses and amelioration of oxidative stress in cadmium (Cd^{2+}) stressed mung bean seedlings: A biochemical and computational analysis. *Journal of Plant Biochemistry and Biotechnology, 31*(1), 49–60. https://doi.org/10.1007/s13562-021-00654-4.

Mazumder, M. K., Sharma, P., Moulick, D., Tata, S., & Choudhury, S., (2021). Salicylic acid ameliorates zinc and chromium induced stress responses in wheat seedlings: A biochemical and computational analysis. *Cereal Research Communications*, 1–12. https://doi.org/10.1007/s42976-021-00201-w.

McLaughlin, M. J., Tiller, K. G., Naidu, R., & Stevens, D. P., (1996). The behavior and environmental impact of contaminants in fertilizers. *Soil Research, 34*(1), 1–54.

Meyer, S., Glaser, B., & Quicker, P., (2011). Technical, economical, and climate-related aspects of biochar production technologies: A literature review. *Environmental Science & Technology, 45*(22), 9473–9483.

Millan, A. F. L., Sagardoy, R., Solanas, M., Abad_ıa, A., & Abadia, J., (2009). Cadmium toxicity in tomato (lycopersicon esculentum) plants grown in hydroponics. *Environmental and Experimental Botany, 65*, 376–385. https://doi.org/10.1016/j.envexpbot.2008.11.010.

Mishra, M., Dashora, K., Srivastava, A., Fasake, V. D., & Nag, R. H., (2019). Prospects, challenges and need for regulation of nanotechnology with special reference to India. *Ecotoxicol. Environ. Saf., 171*, 677–682. https://doi.org/10.1016/j.ecoenv.2018.12.085.

Mitra, S., Singh, P., Manzoor, S., Bhattacharyya, P., Bera, T., Kumar, P. A., Rangan, L., & Borah, P., (2016). Can rice and wheat biochar amendment protect the carbon loss from tropical soils—An experimental study. *Environmental Progress and Sustainable Energy, 35*(1), 183–188.

Moulick, D., Bhutia, K. L., Sarkar, S., Roy, A., Mishra, U. N., Pramanick, B., Maitra, S., et al., (2022b). The intertwining of Zn-finger motifs and abiotic stress tolerance in plants: Current status and future prospects. *Frontiers in Plant Science*. https://doi: 10.3389/fpls.2022.1083960.

Moulick, D., Chowardhara, B., & Panda, S. K., (2019). Agroecotoxicological aspect of arsenic (As) and cadmium (Cd) on field crops and its mitigation: Current status and future prospect. In: *Plant-Metal Interactions* (pp. 217–246). Springer, Cham. https://doi.org/10.1007/978-3-030-20732-8_11.

Moulick, D., Ghosh, D., & Santra, S. C., (2016). Evaluation of effectiveness of seed priming with selenium in rice during germination under arsenic stress. *Plant Physiology and Biochemistry, 109*, 571–578.

Moulick, D., Ghosh, D., & Santra, S. C., (2016a). An assessment of some physicochemical properties and cooking characteristics of milled rice and associated health risk in two rice varieties of arsenic contaminated areas of West Bengal, India. *International Journal of Agricultural and Food Science, 6*, 44–55.

Moulick, D., Ghosh, D., Skalický, M., Gharde, Y., Mazumder, M. K., Choudhury, S., Biswas, J. K., et al., (2022). Interrelationship among rice grain arsenic, micronutrients content and grain quality attributes: An investigation from genotype × environment perspective. *Frontiers in Environmental Science. 10*, 857629 https://doi.org/10.3389/fenvs.2022.857629.

Moulick, D., Hazra, S., Mukherjee, A., Sinha, S., Mahanta, S., Das, A., Saha, B., et al., (2023). Application of nanotechnology in mitigating arsenic stress and accumulation in crops: Where we are and where we are moving towards. In: *Global Arsenic Hazard* (pp. 247–270). Springer, Cham. https://doi.org/10.1007/978-3-031-16360-9_12.

Moulick, D., Samanta, S., Saha, B., Mazumder, M. K., Dogra, S., Panigrahi, K., Banerjee, S., et al., (2020). Salinity stress responses in three popular field crops belonging to fabaceae family: Current status and future prospect. In: *The Plant Family Fabaceae* (pp. 519–541). Springer, Singapore.

Moulick, D., Samanta, S., Sarkar, S., Mukherjee, A., Pattnaik, B. K., Saha, S., & Awashy, J. P., (2021). Arsenic contamination, impact and mitigation strategies in rice agro-environment: An inclusive insight. *Science of the Total Environment, 800*, 149477. https://doi.org/10.1016/j.scitotenv.2021.149477.

Moulick, D., Santra, S. C., & Ghosh, D., (2017). Seed priming with Se alleviate As induced phytotoxicity during germination and seedling growth by restricting As translocation in rice

(*Oryza sativa* L cv IET – 4094). *Ecotoxicology and Environmental Safety, 145*, 449–456. https://doi.org/10.1016/j.ecoenv.2017.07.060.
Moulick, D., Santra, S. C., & Ghosh, D., (2018a). Effect of selenium induced seed priming on arsenic accumulation in rice plant and subsequent transmission in human food chain. *Ecotoxicology and Environmental Safety, 152*, 67–77. https://doi.org/10.1016/j.ecoenv.2018.01.037.
Moulick, D., Santra, S. C., & Ghosh, D., (2018b). Rice seed priming with se: A novel approach to mitigate as induced adverse consequences on growth, yield and as load in brown rice. *Journal of Hazardous Materials, 355*, 187–196. https://doi.org/10.1016/j.jhazmat.2018.05.017.
Moulick, D., Santra, S. C., & Ghosh, D., (2018c). Seed priming with Se mitigates As-induced phytotoxicity in rice seedlings by enhancing essential micronutrient uptake and translocation and reducing as translocation. *Environmental Science and Pollution Research, 25*(27), 26978–26991. https://doi.org/10.1007/s11356-018-2711-x.
Moulick, D., Santra, S. C., & Ghosh, D., (2018d). Consequences of paddy cultivation in arsenic-contaminated paddy fields of lower Indo-Gangetic plain on arsenic accumulation pattern and selected grain quality traits: A preliminary assessment. In: *Mechanisms of Arsenic Toxicity and Tolerance in Plants* (pp. 49–78). Springer, Singapore. https://doi.org/10.1007/978-981-13-1292-2_3.
Moulick, D., Santra, S. C., Ghosh, D., & Panda, S. K., (2019). An assessment of efficiency of zinc priming in rice (cv. MTU-7029) during germination and early seedling growth. In: *Priming and Pretreatment of Seeds and Seedlings* (pp. 495–507). Springer, Singapore. https: Doi.org/10.1007/978-981-13-8625-1_24.
Moulick, D., Sarkar, S., Awasthi, J. P., Ghosh, D., Choudhury, S., Tata, S. K., Bramhachari, K., & Santra, S. C., (2020). Rice grain quality traits: Neglected or less addressed? In: *Rice Research for Quality Improvement: Genomics and Genetic Engineering* (pp. 729–745). Springer, Singapore. https://doi.org/10.1007/978-981-15-4120-9_29.
Mukherjee, A., & Hazra, S., (2022). Impact of elevated CO_2 and O_3 on field crops and adaptive strategies through agro-technology. In: *Response of Field Crops to Abiotic Stress* (pp. 177–190). CRC Press.
Mukherjee, A., Majumdar, S., Servin, A. D., Pagano, L., Dhankher, O. P., & White, J. C., (2016). Carbon nanomaterials in agriculture: A critical review. *Front Plant Sci., 7*, 1–16. https://doi.org/10.3389/fpls.2016.00172.
Munkhbayar, B., Nine, M. J., Jeoun, J., Bat-Erdene, M., Chung, H., & Jeong, H., (2013). Influence of dry and wet ball milling on dispersion characteristics of the multi-walled carbon nanotubes in aqueous solution with and without surfactant. *Powder Technology, 234*, 132–140.
Naghdi, M., Taheran, M., Brar, S. K., Rouissi, T., Verma, M., Surampalli, R. Y., & Valero, J. R., (2017). A green method for production of nanobiochar by ball milling- optimization and characterization. *Journal of Cleaner Production, 164*, 1394–1405.
Naghdi, M., Taheran, M., Brar, S., Kermanshahi-pour, Verma, M., & Surampalli, R. Y., (2017). Immobilized laccase on oxygen functionalized nanobiochars through mineral acids treatment for removal of carbamazepine. *Science of The Total Environment, 584, 585*, 393–401.
Nath, B. K., Chaliha, C., & Kalita, E., (2019). Iron oxide permeated mesoporous rice-husk nanobiochar (IPMN) mediated removal of dissolved arsenic (As): Chemometric modelling and adsorption dynamics. *Journal of Environmental Management, 246*, 397–409.

Nayak, S., Das, K., Bhowmick, M. K., Hossain, S. M., Singh, S., Singh, U. M., & Singh, V. K., (2022). Flood stress prevalence and assessment of farmers' preparedness with resilient rice varieties: A review in context of Eastern India. *Response of Field Crops to Abiotic Stress* (pp. 277–296).

Ngˆo, C., & Van De, V. M. H., (2014). Nanotechnologies in agriculture and food. *Nanotechnol. A Nutshell.* (pp. 233–247). Paris: Atlantis Press. https://doi.org/10.2991/978-94-6239-012-6_13.

Nie, C., Yang, X., Niazi, N. K., Xu, X., Wen, Y., Rinklebe, J., Ok, Y. S., Xu, S., & Wang, H., (2018). Impact of sugarcane bagasse-derived biochar on heavy metal availability and microbial activity: A field study. *Chemosphere, 200*, 274–282.

Nunes, S. M., Paterson, N., Herod, A. A., Dugwell, D. R., & Kandiyoti, R., (2008). Tar formation and destruction in a fixed bed reactor simulating downdraft gasification: Optimization of conditions. *Energy & Fuels, 22*(3), 1955–1964.

Nurfatihah, Z., & Siddiquee, S., (2019). Nanotechnology: Recent trends in food safety, quality and market analysis. In: *Nanotechnol. Appl. Energy, Drug Food* (pp. 283–293). Springer.

Oleszczuk, P., Ćwikła-Bundyra, W., Bogusz, A., Skwarek, E., & Ok, Y. S., (2016). Characterization of nanoparticles of biochars from different biomass. *Journal of Analytical and Applied Pyrolysis, 121*, 165–172.

Osayi, J. I., Iyuke, S., & Ogbeide, S. E., (2014). Biocrude production through pyrolysis of used tyres. *Journal of Catalysts, 2014*, 386371.

Panda, D., Barik, J., & Behera, P. K., (2022). Improving submergence tolerance in rice: Recent progress and future perspectives. In: *Response of Field Crops to Abiotic Stress* (pp. 111–122).

Pang, S., (2019). Advances in thermochemical conversion of woody biomass to energy, fuels and chemicals. *Biotechnology Advances, 37*(4), 589–597.

Pascual, I., Azcona, I., Aguirreolea, J., Morales, F., Corpas, F. J., Palma, J. M., & Sanchez-Diaz, M., (2010). Growth, yield, and fruit quality of pepper plants amended with two sanitized sewage sludges. *Journal of Agricultural And Food Chemistry, 58*(11), 6951–6959.

Perfileva, A. I., Graskova, I. A., Nozhkina, O. A., Zabanova, N. S., Sukhov, B. G., Shkil, N. N., & Nefyodova, E. V., (2019). The current aspects of using chemically synthesized compounds of silver nanoparticles in animal husbandry and agrochemistry. *Nanotechnologies in Russia, 14*(9), 489–496.

Peters, R. J. B., Bouwmeester, H., Gottardo, S., Amenta, V., Arena, M., Brandhoff, P., et al., (2016). Nanomaterials for products and application in agriculture, feed and food. *Trends Food Sci. Technol., 54*, 155–164. https://doi.org/10.1016/j.tifs.2016.06.008.

Plácido, J., Bustamante, L. S., Meissner, K. E., Kelly, D. E., & Kelly, S. L., (2019). Multivariate analysis of biochar-derived carbonaceous nanomaterials for detection of heavy metal ions in aqueous systems. *Science of The Total Environment, 688*, 751–761.

Pradhan, P., (2019). Antagonists to meeting the 2030 agenda. *Nature Sustainability, 2*, 171–172. https://doi.org/10.1038/s41893-019-0248-8.

Pradhan, P., Costa, L., Rybski, D., Lucht, W., & Kropp, J. P., (2017). A systematic study of sustainable development goal (SDG) interactions. *Earth's Future, 5*, 1169–1179. https://doi.org/10.1002/2017EF000632.

Qiao, J. T., Li, X. M., & Li, F. B., (2018). Roles of different active metal-reducing bacteria in arsenic release from arsenic-contaminated paddy soil amended with biochar. *Journal of Hazardous Materials, 344*, 958–967.

Rahman, A., Hossain, M. S., Mahmud, J. A., Nahar, K., Hasanuzzaman, M., & Fujita, M., (2016). Manganese-induced salt stress tolerance in rice seedlings: Regulation of ion homeostasis, antioxidant defense and glyoxalase systems. *Physiol. Mol. Biol. Plants, 22*, 291–306. https://doi.org/10.1007/s12298-016-0371-1.

Ramanayaka, S., Kumar, M., Etampawala, T., & Vithanage, M., (2020). Macro, colloidal and nanobiochar for oxytetracycline removal in synthetic hydrolyzed human urine. *Environmental Pollution, 267*, 115683.

Regon, P., Dey, S., Chowardhara, B., Saha, B., Kar, S., Tanti, B., & Panda, S. K., (2021). Physio-biochemical and molecular assessment of iron (Fe^{2+}) toxicity responses in contrasting indigenous aromatic Joha rice cultivars of Assam, India. *Protoplasma, 258*(2), 289–299.

Rodriguez, O., Peralta-Hernandez, J. M., Goonetilleke, A., & Bandala, E. R., (2019). Biochar-supported nanomaterials for environmental applications. *J. Ind. Eng. Chem., 78*, 21–33. https://doi.org/10.1016/j.jiec.2019.06.008.

Roopan, S. M., & Madhumitha, G., (2018). *Bioorganic Phase in Natural Food: An Overview.* https://doi.org/10.1007/978-3-319-74210-6.

Roy, S., & Rakshit, R., (2022). Impact of climate change on productivity of field crops. In: *Response of Field Crops to Abiotic Stress* (pp. 211–216). CRC Press.

Sagar, L., Praharaj, S., Singh, S., Attri, M., Pramanick, B., Maitra, S., Hossain, A., et al., (2022). Drought and heat stress tolerance in field crops: Consequences and adaptation strategies. In: *Response of Field Crops to Abiotic Stress* (pp. 91–102). CRC Press.

Saha, B., Chowardhara, B., Kar, S., Devi, S. S., Awasthi, J. P., Moulick, D., Tanti, B., & Panda, S. K., (2019). Advances in heavy metal-induced stress alleviation with respect to exogenous amendments in crop plants. In: *Priming and Pretreatment of Seeds and Seedlings* (pp. 313–332). Springer, Singapore.

Sahoo, S., Borgohain, P., Saha, B., Moulick, D., Tanti, B., & Panda, S. K., (2019). Seed priming and seedling pre-treatment induced tolerance to drought and salt stress: Recent advances. *Priming and Pretreatment of Seeds and Seedlings,* 253–263.

Sajjadi, B., Chen, W. Y., Mattern, D. L., Hammer, N., & Dorris, A., (2020). Low-temperature acoustic-based activation of biochar for enhanced removal of heavy metals. *Journal of Water Process Engineering, 34*, 101166.

Saxena, M., Maity, S., & Sarkar, S., (2014). Carbon nanoparticles in 'biochar' boost wheat (*Triticum aestivum*) plant growth. *RSC Advances, 4*(75), 39948–39954.

Shan, Y., Yang, W., Li, Y., Liu, Y., & Pan, J., (2019). Preparation of microwave-activated magnetic bio-char adsorbent and study on removal of elemental mercury from flue gas. *Science of The Total Environment, 697,* 134049.

Smith, P., Bustamante, M., Ahammad, H., Clark, H., Dong, H., Elsiddig, E. A., Haberl, H., et al., (2014). Agriculture, forestry and other land use (AFOLU). In: Edenhofer, O., Pichs-Madruga, R., Sokona, Y., Farahani, E., Kadner, S., Seyboth, K., Adler, A., et al., (eds.), *Climate Change 2014: Mitigation of Climate Change.* Contribution of Working Group III to the Fifth Assessment Report of the Intergovernmental Panel on Climate Change. Cambridge University Press, Cambridge.

Song, B., Cao, X., Gao, W., Aziz, S., Gao, S., Lam, C. H., & Lin, R., (2022). Preparation of nano-biochar from conventional biorefineries for high-value applications. *Renewable and Sustainable Energy Reviews, 157*, 112057.

Song, B., Chen, M., Zhao, L., Qiu, H., & Cao, X., (2019). Physicochemical property and colloidal stability of micron- and nano-particle biochar derived from a variety of feedstock sources. *Science of The Total Environment, 661*, 685–695.

Spokas, K., Koskinen, W., Baker, J., & Reicosky, D., (2009). Impacts of woodchip biochar additions on greenhouse gas production and sorption/degradation of two herbicides in a Minnesota soil. *Chemosphere, 77*(4), 574–581. https://doi.org/10.1016/j.chemosphere.2009.06.053.

Sresty, T. V. S., & Rao, K. M., (1999). Ultrastructural alterations in response to zinc and nickel stress in the root cells of pigeonpea. *Environmental and Experimental Botany, 41*(1), 3–13.

Sridhar, B. B., Witter, J. D., Wu, C., Spongberg, A. L., & Vincent, R. K., (2014). Effect of biosolid amendments on the metal and nutrient uptake and spectral characteristics of five vegetable plants. *Water, Air, and Soil Pollution, 225*(9), 1–14.

Sridhar, B. M., Vincent, R. K., Roberts, S. J., & Czajkowski, K., (2011). Remote sensing of soybean stress as an indicator of chemical concentration of biosolid amended surface soils. *International Journal of Applied Earth Observation and Geoinformation, 13*(4), 676–681.

Sridhar, B. M., Vincent, R. K., Witter, J. D., & Spongberg, A. L., (2009). Mapping the total phosphorus concentration of biosolid amended surface soils using LANDSAT TM data. *Science of the Total Environment, 407*(8), 2894–2899.

Sridhar, M. B., Han, F. X., Diehl, S. V., Monts, D. L., & Su, Y., (2007). Monitoring the effects of arsenic and chromium accumulation in Chinese brake fern *(Pteris vittata)*. *International Journal of Remote Sensing, 28*(5), 1055–1067.

Su, C., (2014). A review on heavy metal contamination in the soil worldwide: Situation, impact and remediation techniques. *Environmental Skeptics and Critics, 3*(2), 24.

Su, Y., Maruthi, S. B. B., Han, F. X., Diehl, S. V., & Monts, D. L., (2007). Effect of bioaccumulation of Cs and Sr natural isotopes on foliar structure and plant spectral reflectance of Indian mustard (*Brassica juncea*). *Water, air, and Soil Pollution, 180*(1), 65–74.

Tomczyk, A., Sokołowska, Z., & Boguta, P., (2020). Biochar physicochemical properties: Pyrolysis temperature and feedstock kind effects. *Reviews in Environmental Science and Bio/Technology, 19*(1), 191–215.

Tóth, G., Hermann, T., Da Silva, M. R., & Montanarella, L. J. E. I., (2016). Heavy metals in agricultural soils of the European Union with implications for food safety. *Environment International, 88*, 299–309.

Tyc, O., Song, C., Dickschat, J. S., Vos, M., & Garbeva, P., (2017). The ecological role of volatile and soluble secondary metabolites produced by soil bacteria. *Trends in Microbiology, 25*(4), pp.280–292.

UN, (2015). *Transforming Our World: The 2030 Agenda for Sustainable Development*. United Nations, New York.

UN, (2022). *World Population Projected to Reach 9.6 Billion by 2050.* https://www.un.org/development/desa/en/news/population/un-report-world-population-projected-to-reach-9-6-billion-by-2050.html (accessed on 25 April 2024).

Wade, S. R., Nunoura, T., & Antal, M. J., (2006). Studies of the flash carbonization process. 2. violent ignition behavior of pressurized packed beds of biomass: A factorial study. *Industrial & Engineering Chemistry Research, 45*(10), 3512–3519.

Wan, X., Li, C., & Parikh, S. J., (2020). Simultaneous removal of arsenic, cadmium, and lead from soil by iron-modified magnetic biochar. *Environmental Pollution, 261*, 114157.

Wang, D., Zhang, W., Hao, X., & Zhou, D., (2013). Transport of biochar particles in saturated granular media: Effects of pyrolysis temperature and particle size. *Environmental Science and Technology, 47*(2), 821–828.
Wang, J., Li, W., Zhao, Z., Musoke, F. S., & Wu, X., (2022). Ultrasonic activated biochar and its removal of harmful substances in environment. *Microorganisms, 10*(8), 1593.
Wang, M. C., Sheng, G. D., & Qiu, Y. P., (2015). A novel manganese-oxide/biochar composite for efficient removal of lead (II) from aqueous solutions. *International Journal of Environmental Science and Technology, 12*(5), 1719–1726.
Wang, M., Zhu, Y., Cheng, L., Andserson, B., Zhao, X., Wang, D., & Ding, A., (2018). Review on utilization of biochar for metal-contaminated soil and sediment remediation. *Journal of Environmental Sciences, 63*, 156–173.
Wang, N., Xue, X. M., Juhasz, A. L., Chang, Z. Z., & Li, H. B., (2017). Biochar increases arsenic release from an anaerobic paddy soil due to enhanced microbial reduction of iron and arsenic. *Environmental Pollution, 220*, 514–522.
Wang, S. Y., Tang, Y. K., Chen, C., Wu, J. T., Huang, Z., Mo, Y. Y., Zhang, K. X., & Chen, J. B., (2015). Regeneration of magnetic biochar derived from eucalyptus leaf residue for lead (II) removal. *Bioresource Technology, 186*, 360–364.
Wang, S., Gao, B., Li, Y., Mosa, A., Zimmerman, A. R., Ma, L. Q., Harris, W. G., & Migliaccio, K. W., (2015). Manganese oxide-modified biochars: Preparation, characterization, and sorption of arsenate and lead. *Bioresource Technology, 181*, 13–17.
Wang, S., Gao, B., Zimmerman, A. R., Li, Y., Ma, L., Harris, W. G., & Migliaccio, K. W., (2015). Removal of arsenic by magnetic biochar prepared from pinewood and natural hematite. *Bioresource Technology, 175*, 391–395.
Wang, Y., Yin, R., & Liu, R., (2014). Characterization of biochar from fast pyrolysis and its effect on chemical properties of the tea garden soil. *Journal of Analytical and Applied Pyrolysis, 110*, 375–381.
Wang, Y., Zhong, B., Shafi, M., Ma, J., Guo, J., Wu, J., Ye, Z., Liu, D., & Jin, H., (2019). Effects of biochar on growth, and heavy metals accumulation of moso bamboo (Phyllostachy pubescens), soil physical properties, and heavy metals solubility in soil. *Chemosphere, 219*, 510–516.
Weber, K., & Quicker, P., (2018). Properties of biochar. *Fuel, 217*, 240–261.
Wei, J., Tu, C., Yuan, G., Liu, Y., Bi, D., Xiao, L., Lu, J., et al., (2019). Assessing the effect of pyrolysis temperature on the molecular properties and copper sorption capacity of a halophyte biochar. *Environmental Pollution, 251*, 56–65.
Wu, J., Li, Z., Huang, D., Liu, X., Tang, C., Parikh, S. J., & Xu, J., (2020). A novel calcium-based magnetic biochar is effective in stabilization of arsenic and cadmium co-contamination in aerobic soils. *Journal of Hazardous Materials, 387*, 122010.
Xiang, Y., Deng, Q., Duan, H., & Guo, Y., (2017). Effects of biochar application on root traits: A meta-analysis. *GCB Bioenergy, 9*, 1563–1572. https://doi.org/10.1111/ gcbb.12449.
Xiao, J., Hu, R., & Chen, G., (2020). Micro-nano-engineered nitrogenous bone biochar developed with a ball-milling technique for high-efficiency removal of aquatic Cd(II), Cu(II) and Pb(II). *Journal of Hazardous Materials, 387*, 121980.
Xiao, R., Wang, S., Li, R., Wang, J. J., & Zhang, Z., (2017). Soil heavy metal contamination and health risks associated with artisanal gold mining in Tongguan, Shaanxi, China. *Ecotoxicology and Environmental Safety, 141*, 17–24.

Xiong, Z. T., & Wang, H., (2005). Copper toxicity and bioaccumulation in Chinese cabbage (*Brassica pekinensis* Rupr.). *Environmental Toxicology: An International Journal, 20*(2), 188–194.

Xu, C. Y., Li, Q. R., Geng, Z. C., Hu, F. N., & Zhao, S. W., (2020). Surface properties and suspension stability of low-temperature pyrolyzed biochar nanoparticles: Effects of solution chemistry and feedstock sources. *Chemosphere, 259*, 127510.

Xu, J., Yang, L., Wang, Z., Dong, G., Huang, J., & Wang, Y., (2006). Toxicity of copper on rice growth and accumulation of copper in rice grain in copper contaminated soil. *Chemosphere, 62*(4), 602–607.

Xu, M., Wu, J., Luo, L., Yang, G., Zhang, X., Peng, H., Yu, X., & Wang, L., (2018). The factors affecting biochar application in restoring heavy metal-polluted soil and its potential applications. *Chemistry and Ecology, 34*(2), 177–197.

Yaashikaa, P. R., Kumar, P. S., Varjani, S., & Saravanan, A., (2020). A critical review on the biochar production techniques, characterization, stability, and applications for circular bioeconomy. *Biotechnology Reports, 28*, e00570.

Yadav, V., & Khare, P., (2020). Impact of pyrolysis techniques on biochar characteristics: Application to soil. *Biochar Applications in Agriculture and Environment Management*, pp. 33–52.

Yang, W., An, H., Chou, S., Chua, K., Mohan, B., Sivasankaralingam, V., Raman, V., Maghbouli, A., & Li, J., (2013). Impact of emulsion fuel with nano-organic additives on the performance of diesel engine. *Applied Energy, 112*, 1206–1212. https://doi.org/10.1016/j.apenergy.2013.02.027.

Ye, Y., Cota-Ruiz, K., Hernandez-Viezcas, J. A., Valdes, C., Medina-Velo, I. A., Turley, R. S., Peralta-Videa, J. R., & Gardea-Torresdey, J. L., (2020). Manganese nanoparticles control salinity-modulated molecular responses in *Capsicum annuum* L. through priming: A sustainable approach for agriculture. *ACS Sustain. Chem. Eng., 8*, 1427–1436. https://doi.org/10.1021/acssuschemeng.9b05615.

Younis, S. A., Kim, K. H., Shaheen, S. M., Antoniadis, V., Tsang, Y. F., Rinklebe, J., Deep, A., & Brown, R. J., (2021). Advancements of nanotechnologies in crop promotion and soil fertility: Benefits, life cycle assessment, and legislation policies. *Renewable and Sustainable Energy Reviews, 152*, 111686.

Yu, Z., Qiu, W., Wang, F., Lei, M., Wang, D., & Song, Z., (2017). Effects of manganese oxide-modified biochar composites on arsenic speciation and accumulation in an indica rice (*Oryza sativa* L.) cultivar. *Chemosphere, 168*, 341–349.

Yu, Z., Zhou, L., Huang, Y., Song, Z., & Qiu, W., (2015). Effects of a manganese oxide-modified biochar composite on adsorption of arsenic in red soil. *Journal of Environmental Management, 163*, 155–162.

Yuan, Y., Zhang, N., & Hu, X., (2020). Effects of wet and dry ball milling on the physicochemical properties of sawdust derived-biochar. *Instrumentation Science & Technology, 48*(3), 287–300.

Yue, L., Lian, F., Han, Y., Bao, Q., Wang, Z., & Xing, B., (2019). The effect of biochar nanoparticles on rice plant growth and the uptake of heavy metals: Implications for agronomic benefits and potential risk. *Science of The Total Environment, 656*, 9–18.

Zahedifar, M., & Najafian, S., (2017). *Ocimum basilicum* L. growth and nutrient status as influenced by biochar and potassium-nano chelate fertilizers. *Archives of Agronomy and Soil Science, 63*(5), 638–650.

Zama, E. F., Reid, B. J., Sun, G. X., Yuan, H. Y., Li, X. M., & Zhu, Y. G., (2018). Silicon (Si) biochar for the mitigation of arsenic (As) bioaccumulation in spinach (*Spinacia oleracean*) and improvement in the plant growth. *J. Clean. Prod., 189*, 386–395. https://doi.org/10.1016/j.jclepro.2018.04.056.

Zhang, C., Shan, B., Zhu, Y., & Tang, W., (2018). Remediation effectiveness of *Phyllostachys pubescens* biochar in reducing the bioavailability and bioaccumulation of metals in sediments. *Environmental Pollution, 242*, 1768–1776.

Zhang, H., Voroney, R. P., Price, G. W., & White, A. J., (2017). Sulfur-enriched biochar as a potential soil amendment and fertilizer. *Soil Res., 55*, 93–99. https://doi.org/10.1071/SR15256.

Zhang, K., & Zhou, Q., (2005). Toxic effects of Al-based coagulants on *Brassica chinensis* and *Raphanus sativus* growing in acid and neutral conditions. *Environmental Toxicology: An International Journal, 20*(2), 179–187.

Zhang, M. M., Liu, Y. G., Li, T. T., Xu, W. H., Zheng, B. H., Tan, X. F., Wang, H., et al., (2015). Chitosan modification of magnetic biochar produced from *Eichhornia crassipes* for enhanced sorption of Cr (VI) from aqueous solution. *RSC Advances, 5*(58), 46955–46964.

Zhang, X., Wells, M., Niazi, N., Bolan, N., Shaheeng, S., Hou, D., Gao, B., et al., (2022). Nanobiochar-rhizosphere interactions: Implications for the remediation of heavy-metal contaminated soils. *Environmental Pollution*, 118810.

Zhang, X., Yan, L., Liu, J., Zhang, Z., & Tan, C., (2019). Removal of different kinds of heavy metals by novel PPG-nZVI beads and their application in simulated stormwater infiltration facility. *Applied Sciences, 9*(20), 4213.

CHAPTER 10

Arsenic Remediation in Rice Through Biochar-Nanoparticle Composites

TRIDIP MONDAL,[1,2] MOHAMMAD SHAHID,[1] RABINDRA KUMAR NAYAK,[2] RUBINA KHANAM,[1] and AMARESH KUMAR NAYAK[1]

[1]*Crop Production Division, ICAR–National Rice Research Institute, Cuttack, Odisha, India*

[2]*Department of Soil Science and Agricultural Chemistry, Odisha University of Agriculture and Technology, Bhubaneswar, Odisha, India*

ABSTRACT

Soil health is the key ingredient for sustainable food production. Healthy soil is a prerequisite to achieving the United Nations' Sustainable Development Goals (SDGs) by 2030. Natural and anthropogenic degradation of soil health, due to heavy metal toxicity, has emerged as a potential threat to global food security and safety. Arsenic (As), a significant geogenic pollutant found ubiquitously on Earth, is one of the major contaminating heavy metals/metalloids. Water and food contaminated with arsenic are considered a global issue, affecting the health of approximately 150 million people worldwide. Humans are primarily exposed to arsenic through water and food, particularly rice. Due to its unique growing ecosystem, submerged/semiaerobic conditions, rice has the ability to accumulate arsenic at levels several times higher than other cereal crops like maize and wheat. Individuals exposed to arsenic are at a higher risk of experiencing both acute and chronic health issues, such as cancer, cardiovascular disease, diabetes, and skin lesions, among others. Several remediation techniques are being employed to address this problem. Biochar, with its high specific surface

Nanomaterials and Nano-Biochar in Reducing Soil Stress: An Integrated Approach to Sustainable Agriculture. Vishnu D. Rajput, & Tatiana Minkina (Eds.)

area, complex porous structure, and various functional groups, is an environmentally sustainable and economically feasible solution. Its modification with nanoparticles can enhance its efficiency in heavy metal remediation. Due to the porous structure of biochar, its fabrication with nanoparticles as a significant supporting material can enhance the stability of nanoparticles, vis-à-vis increasing their arsenic immobilization efficiency using electrostatic interaction, intraparticle diffusion, ligand exchange, complexation, physical adsorption, and co-precipitation mechanisms. Zero-valent iron (ZVI) and silicon dioxide nanoparticle-impregnated rice husk biochar composites have the potential to immobilize arsenic in both paddy growing soil and water systems. It is anticipated that nano ZVI would be more effective than other Fe mineral particles in modifying biochar because of its excellent magnetic property and high reducing ability. In soils, nano ZVI is converted into iron (hydr)oxide and then adsorbs As through the formation of inner-sphere bidentate complexes. Biochar (BC) acts as a semiconductor that facilitates the transformation (corrosion of nano ZVI) in the soil. Silicon dioxide nanoparticle encapsulation also increases the arsenic adsorption capacity of biochar composites and reduces the arsenic uptake and translocation in rice plant parts through the regulation of transporters.

10.1 INTRODUCTION

In drinking water and food crops, particularly rice, arsenic (As) is one of the most challenging environmental issues. It affects millions of people globally, including regions of Southeast Asia, the Bengal Delta, Taiwan, South America, and China (Naidu et al., 2006). Many people rely on groundwater for both their drinking water supply and irrigation water needs for agricultural production, especially rice in arsenic-contaminated areas, which can have significant health effects from long-term arsenic exposure. Arsenic in rice is a major concern for over 3 billion people who consume it as their primary food source, and millions of individuals may be at risk of developing health issues related to it (Meharg & Zhao, 2012). Arsenic exists in various oxidation states, such as As(-III), As(0), As(III), and As(V), depending on the physiochemical conditions in the environment. Arsenite [As(III)] and arsenate [As(V)] are the two main inorganic arsenic species found in natural environments, while mono-, di-, and tri-methyl arsinates are the organic species (Vithanage et al., 2017). Due to its toxicity and potential for human exposure, arsenic has been ranked at the top by the Agency for

Toxic Substances and Disease Registry (ATSDR) (2019). Various levels of arsenic, including concentrations as high as 250,000 mg/kg, can be detected in soils (Mikutta et al., 2014). Arsenic in minute concentrations acts as a causal agent for numerous health issues in humans. Therefore, the World Health Organization (WHO) has maintained a safe and low threshold limit of arsenic at 10 mg/l in water (Vithanage et al., 2017). Rice has a wide range of arsenic levels, with the majority of recorded amounts falling between 0.02 and 0.90 mg/kg (Meharg & Zhao, 2012). Global arsenic speciation and total grain-arsenic concentrations vary widely across rice genotypes. It has been discovered that rice is a significant source of the inorganic form of arsenic, which may account for nearly 10% to 90% of the total arsenic (Williams et al., 2005). Compared to other crops, rice contains around ten times as much arsenic, and rice from countries like India, the US, China, and Bangladesh frequently have considerably higher amounts of inorganic arsenic (Meharg et al., 2008).

In recent years, a great deal of study has been done on several strategies for effectively removing arsenic from soil. However, many of these mostly conventional methods have substantial drawbacks and, to date, have prevented their extensive practical application. These drawbacks include high operating costs, partial As removal, delicate operating conditions, and low efficiency (Sud et al., 2008). Growing interest has been shown in biochar, a carbonaceous substance created by converting agricultural wastes and byproducts, as a consequence of its importance in both agricultural and environmental contexts (Xiang et al., 2019). Numerous research suggests that adding biochar to soil could lessen the bioavailability of inorganic or organic contaminants, improve soil quality, and enhance plant growth (Gao et al., 2019). Additionally, biochar is a popular substance for heavy metal cleanup due to its intricate porous structure, several functional groups, and high specific surface area (Yang et al., 2020). However, because the two anions that were shown to be the most prevalent forms of As in porewater were arsinate (H_2AsO_4, As(V)) and arsinite (H_3AsO_3, As(III)), biochar may only have a limited potential to immobilize As from soil due to its significant negative charge (Du et al., 2019). Therefore, additional research is needed to determine the best ways to modify traditional biochar to improve As removal or immobilization. Various techniques and tools have been used so far to build transformed biochar with improved sorption ability to As (Gao et al., 2020). The sorption effectiveness of biochar can be significantly improved, in particular, by treatment with metal oxyhydroxides (such as aluminum oxide, iron oxide, and manganese oxide). There have been reports of utilizing biochar as soil amendments to lower the amount of As in plant tissues and its bioavailability. Iron, nano zero-valent iron (n-ZVI), and iron-(hydr)oxide are examples of Fe additions that can be used in conjunction with biochar to reduce the uptake of As by reducing As movement and bioavailability (Lebrun et al., 2019; Qiao et al., 2018). Numerous research has been done to improve the sorption ability of biochar by modifying it with nano silica, nano zinc, and other metals.

10.2 PADDY ECOSYSTEM AND ARSENIC

10.2.1 SOURCES OF ARSENIC IN PADDY FIELD

Main causes of arsenic contamination in rice are the use of phosphate fertilizers, waste products from industry, pesticides and herbicides, smelting, coal burning, mining-related activities, and wood preservatives (Jayasumana et al., 2015).

In places where arsenic exists in nature in higher quantities, it may infiltrate wells and other water sources, which can have dangerous effects on the rice growing system as well as on the health of people and animals (Rahman & Hasegawa, 2011). So, the primary causes of arsenic contamination in rice are:

i. the use of As contaminated irrigation water.
ii. soil contaminated with high level of arsenic.
iii. more effectiveness of rice plants in absorbing arsenic from the soil and water compared to other crops.
iv. using As contaminated water at the time of processing or cooking (Islam et al., 2016).

Thus, arsenic has an impact on rice plants, and because they are generally the primary producers, arsenic penetrates the food chain system and has an impact on heterotrophic consumers (cattle and people), who depend on plant products (Zhao et al., 2010).

10.2.2 ARSENIC SPECIATION IN SOIL

An element's speciation is determined by its oxidation states and chemical forms in mineral phases (Mawia et al., 2021). The speciation of a metal conspicuously demonstrates the biological as well as geochemical behavior of that particular metal, despite its total metal level in the soil, while elemental speciation can demonstrate the bioavailability and toxicity level of that element (Pierart et al., 2015). There are several distinct forms of arsenic in the soil, including the adsorbed form, exchangeable form, free As (ionic form), precipitated solid form, and firmly attached to other different minerals. In a submerged rice field, arsenous acid (AsIII, $As(OH)_3$) is the predominating As species, which has a similar molecular size to silicic acid ($Si (OH)_4$) and pKa (H_4SiO_4, diameter 4.38 Å, pKa = 9.8; H_3AsO_3, diameter 4.11 Å, pKa = 9.2) (Zhang et al., 2020). It is believed that the As component that forms the structural basis for other minerals is not bioavailable. However, the extractable form of As that is adsorbed on minerals or soil is extremely dangerous for the living community (Li et al., 2015). Mobility, bioavailability, absorption, and toxicity of As in soil-plant systems are all governed by its oxidation state. In the natural environment, arsenic has been found in two significant oxidation states, such as As(V) and As(III). Additionally, the different types of oxidation states of As in the soil are greatly altered by various environmental factors. Dimethylarsinate (DMA(V)), Monomethylarsonate (MMA(V)), arsenosugars, arsenobetaine, arsenocholine, monomethylarsonite (MMA(III)), dimethylarseinite (DMA(III)), and trimethylarsine oxide, tetramethylarsonium ions are the other As species found in various soils (Joseph et al., 2015). $HAsO_4^{2-}$ and $H_2AsO_4^-$ are the main inorganic and stable forms of As(V), while H_3AsO_3 is for As(III). Among these, $H_2AsO_4^-$ is the predominant As species found in aerobic soils and is extensively absorbed by plants.

According to their solubility, the As species are normally bioavailable in the following order to rice plants: MMA > AsIII > AsV > DMA. With trace amounts of MMA, DMA, and iAs (inorganic-As) are the predominant As species in rice grains (Wu et al., 2017). As(III), which predominates in anoxic paddy soils, makes up the majority of the iAs in rice grains.

The soil pH, which controls the speciation of metals, is one of the most crucial elements. The desorption process of As from iron oxides is accelerated by higher pH (pH 8.5), which might augment the negative surface charges, specifically hydroxyl ions. Arsenic is then mobilized into the soil solution and may be absorbed by plants in significant amounts. The speciation of As in the soil is also significantly impacted by the redox condition of the soil. Because As is so sensitive to the redox state of the soil, lowering the soil redox potential increases As's availability and solubility in the soil (Gorny et al., 2015). Some studies suggest that the less hazardous As(V) can persist in the soil under oxidizing circumstances. As(V) is easily transformed into As(III), a highly poisonous form, in reduced soil conditions (Khalid et al., 2017). Similar to how soil pH is impacted, microorganisms also have an impact on the speciation and availability of metals in the soil-plant system. Microorganisms have a significant impact on the desorption, adsorption, mobility, solubility, bioavailability, and translocation of arsenic into the soil and plant system. By changing the speciation and interconverting As(III) and As(V), microbes may play a significant role in limiting the transfer of As to plants and have a positive impact on remediation. Numerous investigations have identified anaerobic and aerobic bacteria that actively contribute to the conversion of As species. These bacteria were recovered from As-contaminated areas. *Geospirillum arsenophilus, Geospirillum barnesi, Bacillus arsenicoselenatis, P. arsenitoxidans, Crysiogenes arsinates, Desulfutomaculum auripigmentu, Thermus thermophiles, Thermus aquaticus are* some of the bacterial species that actively contribute to conversion of As(III) into As(V) and the production of arsinite oxidase. However, some bacterial species also participate in the reduction process that turns As(V) into As(III), and these important microorganisms are *Desulfotomaculum auripigmentum, Sulfurospirillum arsenophilum, Geospirillum arsenophilus, Bacillus arsenicus, Crysiogenes arsenatis*, and *Sulfurospirillum barnesii.*

10.2.3 ARSENIC UPTAKE AND TRANSLOCATION IN RICE PLANT

The adsorption and bioavailability of As to root secretions, as well as microbial symbiotic interactions with the root system, are two important elements that significantly affect the arsenic uptake from the soil solution to the roots. However, there are primarily three ways for As to be transferred from the soil to the roots. First, active transport via the symplastic pathway; second, passive transport via the apoplastic pathway; and third, intercellular transport via the plant vascular system. Different transporters and channels allow plant roots to selectively absorb various kinds of As. Arsinate (As(V)), which is common in aerobic soils and has a chemical resemblance to phosphates, can enter plant root cells via phosphate transporters. Arsenite (As(III)), which is the predominant form of arsenic in anaerobic soil environments like submerged paddy soils, has been discovered to be absorbed by plants through aquaporins and membrane channels, which transport neutral molecules and water (Li et al., 2014; Mukhopadhyay et al., 2014). There are numerous members of the nodulin 26-like intrinsic protein (NIP) family, such as NIP2, NIP3, NIP5, NIP1, NIP6, and NIP7, that actively contribute to the uptake and translocation of As(III) in a number of plant species. In addition, the uptake of As(III) by roots has been linked to the silicon transporter (OsLsi1) in rice, which is present in both the exodermis and the endodermis and suitable for As entrance into rice. OsLsi2 and OsLsi3, on the other hand, have reportedly been implicated in transporting As from rice roots to certain xylem vessels, where it is then transferred to rice grains. The second protein group known as plasma membrane intrinsic protein (PIP) members (OsPIP2; 6, OsPIP2; 7, and OsPIP2; 4) enables the entrance and transfer of As(III) in numerous plant species by displaying downregulation in transcript level. The possible mechanism of uptake of As and translocation in rice is depicted in Figure 10.1 (Ali et al., 2009).

Due to anthropogenic activities, including the application of fungicides, herbicides, and pesticides containing arsenic, as well as the biomethylation of arsenic by microorganisms, methylated forms of arsenic, such as dimethylarsinic acid (DMA) and monomethylarsinic acid (MMA), may be deposited. Methylated forms of arsenic are produced by soil and rhizospheric bacteria and are also present in higher concentrations in several plant organs. Rice grain has twice as much methylated forms of arsenic as inorganic arsenic (Zheng et al., 2013). However, despite the fact that the precise mechanism underpinning the uptake and mobilization of methylated arsenic species by plants is not fully described, additional investigations demonstrate the inward entry of DMA and MMA by rice (OsLsi1) transporter. Additionally, it has been noted that rice plants lacking OsLsi1 have 50% to 80% lower absorption of arsenic (MMA and MDA) than wild plants growing in an arsenic-stressed environment. Rice grains absorb the methylated arsenic species extremely effectively, while roots do not absorb as much of it. This might be a key factor contributing to rice spikelet disorder.

It is poorly understood how As is absorbed and transported within the rice plant system. The uptake of As by rice is greatly affected by the levels of Si and P in pore water and soil. Due to the analogous nature of As(V) with phosphate, two primary absorption mechanisms have been discovered in rice. The undissociated methylated DMA and MMA, as well as the silica analog As(III), may also enter the root through different types of aquaporin channels, which are used for silicate transport, according to recent studies. The uptake effectiveness of various As species varies significantly. Although methylated As species have a lower uptake efficiency than inorganic As species, they have a higher translocation efficiency. According to studies, rice grains transfer more As than other types of cereals. This results from the variation in the transfer ratio, and the high soil-to-shoot ratio is most likely caused by the As speciation and dynamics that differ between anaerobic and aerobic soils (Figure 10.1).

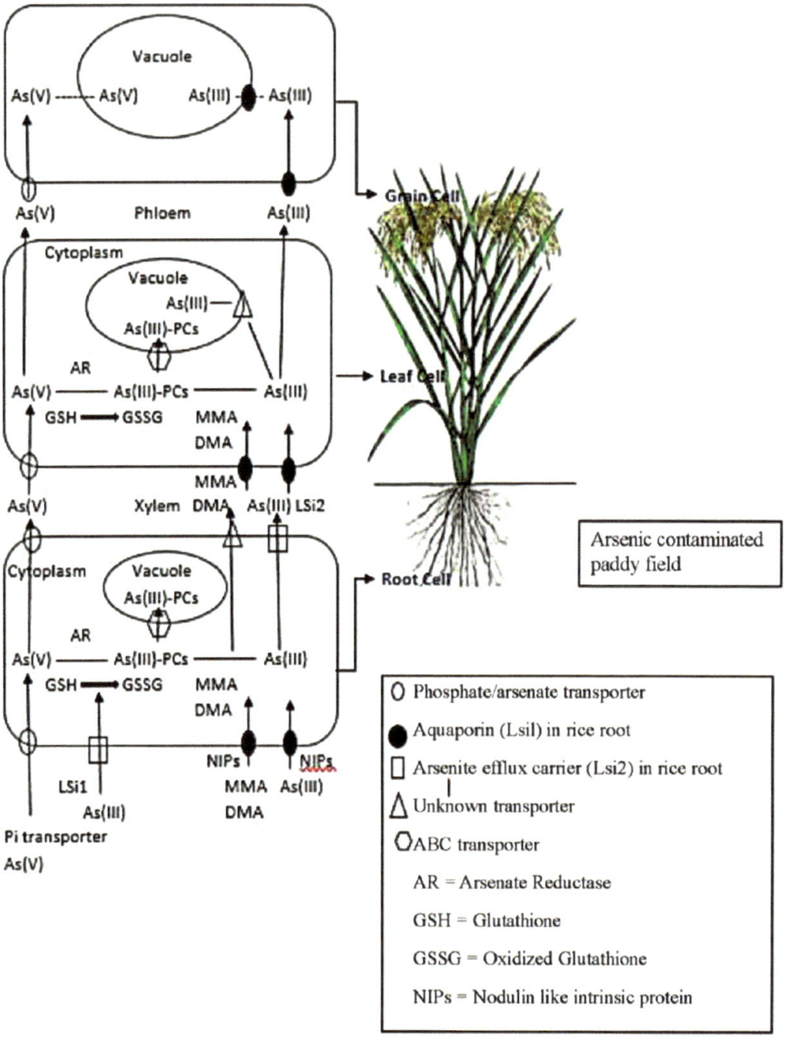

FIGURE 10.1 Conceptual framework of arsenic (As) uptake and mobilization in rice.
Source: Adapted from Ali et al., 2009

10.2.4 PHYTOTOXICITY OF ARSENIC IN RICE

Arsenic is a severely phytotoxic chemical, and rice shows non-tolerance to high soil arsenic concentrations, which lowers plant development and crop output. Many researchers have looked at how As impacts rice growth, and the vast majority of researchers have demonstrated its adverse effect on rice growth, development, and yield. It has been shown that when the rice crops were irrigated with water contaminated with As, plant height, number of tillers, shoot biomass, and grain and straw yield significantly reduced.

The straight head disease affects rice physiologically and causes delayed heading dates, shorter plant heights, and drastically lower grain production. It has been associated with As (Gilmour & Wells, 1980). With an increase in the concentration of soil As, the severity of straight head considerably worsens. Straight head can reduce grain yield by 16–100% and cause the generation of around 17–100% sterile spikelets. Chlorophyll-a and Chlorophyll-b, two photosynthetic pigments that are closely related to rice yield and growth, are reportedly affected by As. When soil As concentration goes up to 30 mg kg^{-1}, both the chlorophyll-a and chlorophyll-b levels in rice leaves considerably decrease.

10.3 ARSENIC CONSUMPTION THROUGH RICE AND HUMAN HEALTH EXPOSURE

Although rice has been recognized as a substantial exposure channel for daily As consumption, drinking water has long been thought to be the primary exposure pathway for As. For the Asian population, rice can make up a significant portion of their daily diet, up to 67% of it. Numerous nations have recently released findings from dietary exposure research. There are considerable differences in the estimated daily intake between nations. The average daily consumption of rice in Asian nations is 588 g day^{-1} (FAOSTAT, 2007), and compared to other foods, it contributes a comparatively significant amount. According to EFSA, rice makes up 5% to 15% of the diet's inorganic As consumption, depending on the age group (EFSA, 2014). According to Mondal et al. (2010), rice is the predominant source of inorganic As exposure in many blocks of West Bengal, India, and may account for 12 to 34% of all As. For the adult population, the regulation establishes upper limits of 0.2 mg As kg^{-1} for white rice and 0.25 mg As kg^{-1} for brown rice.

The results of current studies demonstrate that exposure to As has negative impacts on human health. Being exposed to arsenic in high concentrations can lead to significant illnesses, including various malignancies (IARC, 2004). Generally, a mixture of As metabolites discharged in urine at relatively steady amounts includes 60–70% DMA, 10–30% inorganic As, and 10–20% MMA, and it is linked to a number of internal cancers, including those of the kidney, lungs, liver, and bladder, as well as other health issues like diabetes and skin cancer. Inorganic As is largely removed through the urine and is metabolized in the liver through sequential reduction and oxidative methylation. This mechanism is believed to function as a detoxification mechanism since the MMA, DMA, and primary methylation metabolites are swiftly removed and less critically dangerous than the inorganic species. A clinical study of 86,000 people from West Bengal (India) and 18,000 people from Bangladesh who lived in As-affected areas revealed that 3,695 of these people (20.6% of whom were children) in Bangladesh and 8,500 of these people (9.8% of whom were kids) in West Bengal (India) had arsenic-related skin diseases (Rahman et al., 2001). A significant part of the population is incredibly susceptible to a number of internal cancers. However, the length of time it takes to develop arsenicosis varies depending on many factors such as the amount of arsenic (As) in food and drink, a person's nutritional state, their genetic make-up, and other aggravating factors. In urothelial cell micronuclei, it has been shown that cooked rice with 0.20 mg As kg^{-1} exerts genotoxic effects on human beings (Banerjee et al., 2013). The health risk index (HRI) and chronic daily intake (CDI) for rice were found to be high, according to an Indian study, indicating possible health-related problems in the Indian middle Gangetic plains (Kumar et al., 2016). Numerous studies have shown that eating rice increases your lifetime chance of developing cancer.

10.4 REMEDIATION OF ARSENIC THROUGH BIOCHAR-NANOPARTICLE COMPOSITES

It has been widely reported that biochar is utilized as a soil ameliorant to lessen the bioavailability of metals and metalloids (Zhang et al., 2013). Because As is only marginally retained on the surface of biochar, it can lower the As content in plant tissues. The following are some hypothesized processes for heavy metal stabilization after applying biochar to polluted soils (Zhang et al., 2013):

(i) Ion Exchange: Metal ions present in the soil can exchange with Mg^{2+}, Ca^{2+}, and other cations associated with biochar to create surface innersphere complexes or co-precipitation with biochar since biochar often has a high pH and capacity for cation exchange.
(ii) Ligand complexation: Functional groups on biochar, such as phenolic hydroxyl groups, carboxyl, or alcoholic hydroxyl, can react in a complexation reaction with heavy metals/metalloids through inner-sphere-specific surface complexations and electrostatic attraction.
(iii) Surface precipitation and physical or chemical sorption.

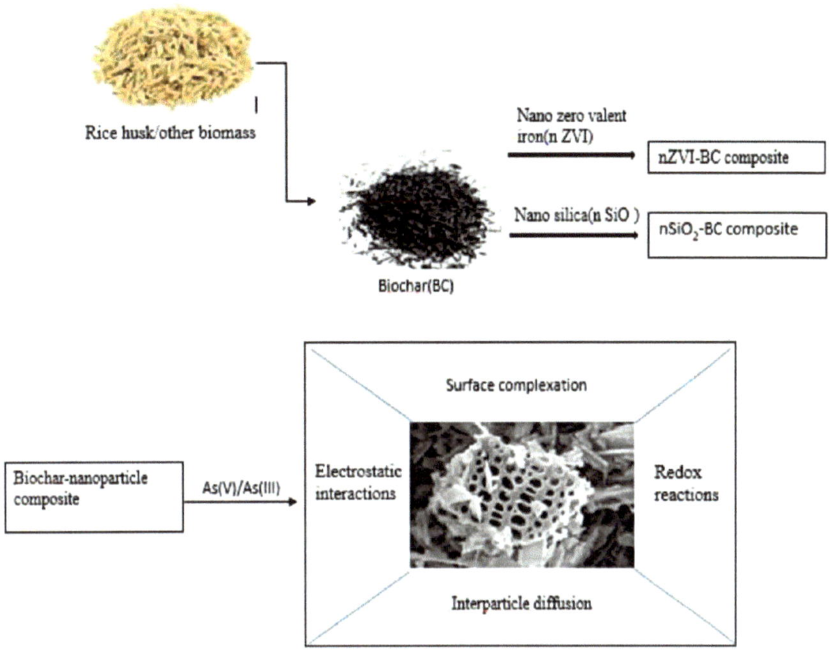

FIGURE 10.2 Simple diagram of biochar-nanoparticle composite preparation and its interaction with arsenic.

Most biochars are ineffective at eliminating anionic contaminants, such as arsenic. The selectivity and removal efficiency of biochar may therefore be improved by modifying its hydrophobicity/hydrophilicity, ash concentrations, functional groups, surface charges, and surface area through chemical, physical, thermal, and/or mechanical processes (Ahmad et al., 2018). Due to the nano-size, huge surface area, and reductive capabilities of nano zero valent iron (nZVI), combining biochar with ZVI may, for instance, result in the development of an effective magnetic adsorbent. A simple picture of the biochar-nanoparticle interaction with arsenic is given in Figure 10.2.

10.4.1 ARSENIC REMEDIATION MECHANISM OF NANO ZERO-VALENT IRON (NZVI)-BIOCHAR COMPOSITES

According to a recent investigation, Fe supplements can lower As uptake in paddy rice, including the amorphous form of iron-(hydr)oxide, ferrous form of iron, and iron powder. This is due to the fact that an increase in soil iron (hydr)oxides might control the mobility and bioavailability of As through reductive dissolution or mineralization processes. Zero-valent iron (ZVI) has proven to play distinct roles in immobilizing As in contaminated soil, thanks to the significant incorporation and absorption capacities of its corrosion products for As. Recently, there has been an increased emphasis on the discovery that zero-valent iron (ZVI), particularly nanoscale ZVI (nZVI), is more effective at stabilizing As than iron complexes (Dong et al., 2019). This is because of the remarkable magnetic property and exceptional As(V) removal capacity of nZVI, which are expected to make it more efficient than other Fe/mineral particles, such as natural hematite, maghemite, birnessite (MnO_2), manganese ferrite ($MnFe_2O_4$), and aluminum oxide (Al_2O_3), in modifying biochars (Wang et al., 2016). Inner-sphere bidentate binuclear complexes are produced as a result of nZVI's conversion to iron (hydr)oxide in soils, and they eventually adsorb As. Although nZVI showed efficient As immobilization, its stability in soil is troublesome due to the high surface energy that causes it to clump and consequently lose its sorption capability (Mandal et al., 2020).

However, when a composite is created using biochar (BC), this biochar accelerates the corrosion of nZVI in the soil because of its capacity for electrical conductivity and electron donation (Prevoteau et al., 2016). Biochar may serve as an "electronic shuttle" because, when nZVI is attached to the surface of biochar, electron transport between the two occurs directly rather than through simple combination. In this way, the arsenic remediation efficiency of biochar nanoparticle composites is increased compared to biochar and nanoparticles alone.

When a ZVI-BC composite is applied to soil, a couple of redox reactions occur in which As^{+5} is reduced to As^{+3} and Fe° is first oxidized to Fe^{+2} and subsequently to Fe^{+3}. Additionally, we conclude that As^{+3} and Fe^{+3} co-precipitate with the creation of distinct Fe(As)OOH phases.

The Scanning Electron Micrograph (SEM) depicts what might be a thin oxyhydroxide surface layer coating of Fe(As)OOH on top of a biochar particle and a distinct remnant Fe° particle. Evidence suggests that as ZVI particles are dissolved by corrosion processes, the unique Fe(As)OOH phases are neo-formed on nearby biochar surfaces. The biochar, a semiconductor, may have aided in the transmission of electrons from the ZVI to O_2 or As^{+3} ions, promoting the precipitation reaction. In order to reduce As^{+5} to As^{+3}, ZVI supplies an electron, and As^{+3} co-precipitates with Fe^{+3} to create secondary Fe(As)- oxyhydroxides on biochar surfaces.

10.4.2 ROLE OF NANO SILICON DIOXIDE (N SIO_2)-BIOCHAR COMPOSITES IN ARSENIC REMEDIATION

Due to the high specific surface area of SiO_2 and better hydrothermal stability, creating a composite of biochar and SiO_2 may enhance the physicochemical properties of the material, improving its As(V) adsorption capability. Additionally, the produced composite of SiO_2 and biochar may also help fix other organic and inorganic contaminants (Ahmad et al., 2018). It has been reported that mesoporous silica sorbents can adsorb As(V) through electrostatic action (Fan et al., 2012). After the inclusion of Si-enriched biochar composites in paddy soil, several crucial processes lead to the immobilization of As through different steps:

i. Reductive dissolution of Fe(III)(oxyhydr)oxides releases Fe(II).
ii. Silicate (SiO_{4-}^{4} as well as SiO_{3-}^{3}) adsorption of Fe(II) and its oxidation to Fe(III) on biochar surface.
iii. At a pH close to neutral, ferrihydrite layer development on the biochar surface is triggered by the complexation of Fe(III) with the hydroxide ion.
iv. Production of As(V) through As(III) oxidation by microbes and subsequent adsorption on BC-Si-Fe(III)OOH.

There are several stages involved in the arsenic uptake by rice plants in the presence of Si-modified rice husk biochar material. The primary mechanism in rice roots that controls As(III) uptake is called the Si-uptake pathway and it includes: (i) the rhizosphere receiving labile Si released from biochar, (ii) reducing the number of transporters accessible for As(III) absorption by downregulating Si-transporters in roots under high Si levels, and (iii) As(III) absorption by rice is reduced as a result of increased competition between Si and As(III) (Seyfferth et al., 2016).

10.5 STUDY OF LONG TERM/PERMANENT STABILITY OF IMMOBILIZED ARSENIC IN SOIL

Soil has to be taken at a depth of 0-30 cm from the As-contaminated area. Using a 60 cm long PVC column with a 19 cm diameter, deionized water is leached through the arsenic-polluted soil to create arsenic-laden groundwater. Rough quartz sand, with a particle size of 2–3 mm, is packed into the bottom of the column to a height of approximately 12 cm. The contaminated soil, weighing roughly 7.2 kg, is layered 20 cm above the coarse quartz sand. The soil in the column has a density of 1.27 g/cm³. The effluent solution is then collected to represent contaminated groundwater after a specific volume of deionized water has been applied to the soil. To maintain a consistent ratio of As-contaminated soil to groundwater, which is 1:2.5 (w/v), the volume of arsenic-contaminated groundwater is set at 36 liters for each column, with the amount of contaminated soil in each column weighing approximately 14.4 kg (including 7.2 kg of topsoil and 7.2 kg of subsoil, as discussed in the following paragraph). The remediation agents for As immobilization are nZVI-BC and $nSiO_2$-BC. The demonstration test of simultaneous soil and groundwater remediation is conducted using the column for polluted groundwater preparation described in the previous paragraph. The soil that remains in the column after filtration is kept as subsurface because the site investigation revealed that the subsoil's As content is significantly lower than that of the topsoil. The topsoil is then layered 20 cm above the subsoil using 7.2 kg of the arsenic-polluted soil mixed with 5% (w/w) of nZVI-BC and n SiO_2-BC separately. The control is topsoil devoid of nZVI-BC and n SiO_2-BC addition. Then, in an unsaturated environment, a peristaltic pump continuously injects 36.0 L of polluted groundwater at a flow rate of 18.0 L d^{-1} into the top of the soil. The effluent water also returns to the topsoil at the same time. In this approach, a cyclic system for integrated soil and groundwater remediation is constructed, with one cycle having a duration of two days. After the start of the system, approximately 10 mL of influent and effluent water are collected on a regular basis, and As concentrations are monitored. The experiment continues until the total As concentration of influent and effluent waters is almost equal, indicating that the system has reached equilibrium. At the conclusion of the experiment, soil samples are taken from the top to the bottom of the columns at 0–5, 5–10, 10–15, 15–20, 20–25, 25–30, 30–35, and 35–40 cm layers, respectively. These soil samples undergo immobilization examination utilizing the toxicity characteristics leaching procedure (TCLP) test after being air-dried (USEPA, 2004). The effectiveness of remediation technologies for contaminated soils, as well as the leaching potential of contaminants under simulated landfill conditions, has both been thoroughly assessed using the TCLP technique (Tsang et al., 2013).

10.6 CONCLUSION

The efficacy of arsenic immobilization in contaminated soil is directly impacted by the alteration of the biochar surface using various forms of Si, especially nano silica ($nSiO_2$) and nano zero valent iron (nZVI). Under a variety of environmental conditions, the nZVI-BC and $nSiO_2$-BC composites exhibit outstanding performance in the stabilization of As in soil. Due to the sturdy binding of heavy metals/metalloids following the application of nZVI-BC and $nSiO_2$-BC composites, the leaching of heavy metals/metalloids can be minimized. By means of precipitation/co-precipitation, reduction, complexation, and the creation of ternary surface complexes, nZVI-BC and $nSiO_2$-BC composites immobilize arsenic. In comparison to biochar and nanoparticles alone, nZVI-BC and $nSiO_2$-BC composites have a wider potential for the remediation of heavy metals/metalloids in arsenic-affected soil due to their adaptability and high efficiency.

KEYWORDS

- **arsenic**
- **biochar-nano composites**
- **silicon dioxide nanoparticle**
- **zero valent iron**

REFERENCES

Ahmad, M., Ahmad, M., Usman, A. R., Al-Faraj, A. S., Abduljabbar, A. S., & Al-Wabel, M. I., (2018). Biochar composites with nano zerovalent iron and eggshell powder for nitrate removal from aqueous solution with coexisting chloride ions. *Environmental Science and Pollution Research, 25*(26), 25757–25771.

Ali, W., Isayenkov, S. V., Zhao, F. J., & Maathuis, F. J., (2009). Arsinite transport in plants. *Cellular and Molecular Life Sciences, 66*(14), 2329–2339.

Banerjee, M., Banerjee, N., Bhattacharjee, P., Mondal, D., Lythgoe, P. R., Martínez, M., & Giri, A. K., (2013). High arsenic in rice is associated with elevated genotoxic effects in humans. *Scientific Reports, 3*(1), 1–8.

Barrachina, A. C., Carbonell, F. B., & Beneyto, J. M., (1995). Arsenic uptake, distribution, and accumulation in tomato plants: Effect of arsinite on plant growth and yield. *Journal of Plant Nutrition, 18*(6), 1237–1250.

Correll, R., Huq, S. I., Smith, E., Owens, G., & Naidu, R., (2006). Dietary intake of arsenic from crops. *Managing Arsenic in the Environment.*

Dong, H., Li, L., Lu, Y., Cheng, Y., Wang, Y., Ning, Q., & Zeng, G., (2019). Integration of nanoscale zero-valent iron and functional anaerobic bacteria for groundwater remediation: A review. *Environment International, 124*, 265–277.

Du, F., Wang, L., Yang, Z., Liu, P., & Li, D., (2019). Ionomic profile and arsenic speciation in *Semisulcospira cancellata*, a freshwater shellfish from a mine-impacted river in China. *Environmental Science and Pollution Research, 26*(10), 10148–10158.

European Food Safety Authority, (2014). Dietary exposure to inorganic arsenic in the European population. *EFSA Journal, 12*(3), 3597.

Fan, H. T., Fan, X., Li, J., Guo, M., Zhang, D., Yan, F., & Sun, T., (2012). Selective removal of arsenic (V) from aqueous solution using a surface-ion-imprinted amine-functionalized silica gel sorbent. *Industrial & Engineering Chemistry Research, 51*(14), 5216–5223.

FAOSTAT, (2007*). Food Consumption.* Food and Agriculture Organization of the United Nations.

Gao, X., Peng, Y., Guo, L., Wang, Q., Guan, C. Y., Yang, F., & Chen, Q., (2020). Arsenic adsorption on layered double hydroxides biochars and their amended red and calcareous soils. *Journal of Environmental Management, 271*, 111045.

Gao, X., Peng, Y., Zhou, Y., Adeel, M., & Chen, Q., (2019). Effects of magnesium ferrite biochar on the cadmium passivation in acidic soil and bioavailability for packoi (*Brassica chinensis* L.). *Journal of Environmental Management, 251*, 109610.

Gilmour, J. T., & Wells, B. R., (1980). Residual effects of MSMA on sterility in rice cultivars 1. *Agronomy Journal, 72*(6), 1066, 1067.

Gorny, J., Billon, G., Lesven, L., Dumoulin, D., Madé, B., & Noiriel, C., (2015). Arsenic behavior in river sediments under redox gradient: A review. *Science of the Total Environment, 505*, 423–434.

IARC Working Group on the Evaluation of Carcinogenic Risks to Humans, World Health Organization, & International Agency for Research on Cancer, (2004). *Some Drinking-Water Disinfectants and Contaminants, Including Arsenic* (Vol. 84). IARC.

Islam, S., Rahman, M. M., Islam, M. R., & Naidu, R., (2016). Arsenic accumulation in rice: Consequences of rice genotypes and management practices to reduce human health risk. *Environment International, 96*, 139–155.

Jayasumana, C., Fonseka, S., Fernando, A., Jayalath, K., Amarasinghe, M., Siribaddana, S., & Paranagama, P., (2015). Phosphate fertilizer is a main source of arsenic in areas affected with chronic kidney disease of unknown etiology in Sri Lanka. *Springer Plus, 4*(1), 1–8.

Joseph, T., Dubey, B., & McBean, E. A., (2015). Human health risk assessment from arsenic exposures in Bangladesh. *Science of the Total Environment, 527*, 552–560.

Khalid, S., Shahid, M., Niazi, N. K., Rafiq, M., Bakhat, H. F., Imran, M., & Dumat, C., (2017). Arsenic behavior in soil-plant system: Biogeochemical reactions and chemical speciation influences. In: *Enhancing Cleanup of Environmental Pollutants* (pp. 97–140). Springer, Cham.

Kumar, M., Rahman, M. M., Ramanathan, A. L., & Naidu, R., (2016). Arsenic and other elements in drinking water and dietary components from the middle Gangetic plain of Bihar, India: Health risk index. *Science of the Total Environment, 539*, 125–134.

Lebrun, M., Miard, F., Nandillon, R., Scippa, G. S., Bourgerie, S., & Morabito, D., (2019). Biochar effect associated with compost and iron to promote Pb and As soil stabilization and *Salix viminalis* L. growth. *Chemosphere, 222*, 810–822.

Li, G., Santoni, V., & Maurel, C., (2014). Plant aquaporins: Roles in plant physiology. *Biochimica et Biophysica Acta (BBA)-General Subjects, 1840*(5), 1574–1582.

Li, H. B., Li, J., Zhu, Y. G., Juhasz, A. L., & Ma, L. Q., (2015). Comparison of arsenic bioaccessibility in housedust and contaminated soils based on four in vitro assays. *Science of the Total Environment, 532*, 803–811.

Mandal, S., Pu, S., Shangguan, L., Liu, S., Ma, H., Adhikari, S., & Hou, D., (2020). Synergistic construction of green tea biochar supported nZVI for immobilization of lead in soil: A mechanistic investigation. *Environment International*, *135*, 105374.

Mawia, A. M., Hui, S., Zhou, L., Li, H., Tabassum, J., Lai, C., & Hu, P., (2021). Inorganic arsenic toxicity and alleviation strategies in rice. *Journal of Hazardous Materials*, *408*, 124751.

Meharg, A. A., & Zhao, F. J., (2012). Arsenic in rice grain. In: *Arsenic & Rice* (pp. 11–30). Springer, Dordrecht.

Meharg, A. A., Lombi, E., Williams, P. N., Scheckel, K. G., Feldmann, J., Raab, A., & Islam, R., (2008). Speciation and localization of arsenic in white and brown rice grains. *Environmental Science & Technology*, *42*(4), 1051–1057.

Mikutta, C., Mandaliev, P. N., Mahler, N., Kotsev, T., & Kretzschmar, R., (2014). Bioaccessibility of arsenic in mining-impacted circumneutral river floodplain soils. *Environmental Science & Technology*, *48*(22), 13468–13477.

Mondal, D., Banerjee, M., Kundu, M., Banerjee, N., Bhattacharya, U., Giri, A. K., & Polya, D. A., (2010). Comparison of drinking water, raw rice and cooking of rice as arsenic exposure routes in three contrasting areas of West Bengal, India. *Environmental Geochemistry and Health*, *32*(6), 463–477.

Mukhopadhyay, R., Bhattacharjee, H., & Rosen, B. P., (2014). Aquaglyceroporins: Generalized metalloid channels. *Biochimica et Biophysica Acta (BBA)-General Subjects*, *1840*(5), 1583–1591.

Naidu, R., Smith, E., Owens, G., & Bhattacharya, P., (2006). *Managing Arsenic in the Environment: From Soil to Human Health*. CSIRO publishing.

Nordstrom, D. K., (2002). Worldwide occurrences of arsenic in ground water. *Science*, *296*(5576), 2143–2145.

Pierart, A., Shahid, M., Sejalon-Delmas, N., & Dumat, C., (2015). Antimony bioavailability: Knowledge and research perspectives for sustainable agricultures. *Journal of Hazardous Materials*, *289*, 219–234.

Prévoteau, A., Ronsse, F., Cid, I., Boeckx, P., & Rabaey, K., (2016). The electron donating capacity of biochar is dramatically underestimated. *Scientific Reports*, *6*(1), 1–11.

Qiao, J. T., Liu, T. X., Wang, X. Q., Li, F. B., Lv, Y. H., Cui, J. H., & Liu, C. P., (2018). Simultaneous alleviation of cadmium and arsenic accumulation in rice by applying zero-valent iron and biochar to contaminated paddy soils. *Chemosphere*, *195*, 260–271.

Rahman, M. A., & Hasegawa, H., (2011). High levels of inorganic arsenic in rice in areas where arsenic-contaminated water is used for irrigation and cooking. *Science of the Total Environment*, *409*(22), 4645–4655.

Rahman, M. M., Chowdhury, U. K., Mukherjee, S. C., Mondal, B. K., Paul, K., Lodh, D., & Chakraborti, D., (2001). Chronic arsenic toxicity in Bangladesh and West Bengal, India—A review and commentary. *Journal of Toxicology: Clinical Toxicology*, *39*(7), 683–700.

Seyfferth, A. L., Morris, A. H., Gill, R., Kearns, K. A., Mann, J. N., Paukett, M., & Leskanic, C., (2016). Soil incorporation of silica-rich rice husk decreases inorganic arsenic in rice grain. *Journal of Agricultural and Food Chemistry*, *64*(19), 3760–3766.

Smedley, P., & Kinniburgh, D., (2002). A review of the source, behavior and distribution of arsenic in natural waters. *Appl. Geochem.*, *17*, 517–568.

Sud, D., Mahajan, G., & Kaur, M. P., (2008). Agricultural waste material as potential adsorbent for sequestering heavy metal ions from aqueous solutions–A review. *Bioresource Technology*, *99*(14), 6017–6027.

Tsang, D. C., Olds, W. E., Weber, P. A., & Yip, A. C., (2013). Soil stabilization using AMD sludge, compost and lignite: TCLP leachability and continuous acid leaching. *Chemosphere, 93*(11), 2839–2847.

USEPA, (2004). *OSWER, ORCR, SW-846 Test Method 1311: Toxicity Characteristic Leaching Procedure* (pp. 1–35).

Vithanage, M., Herath, I., Joseph, S., Bundschuh, J., Bolan, N., Ok, Y. S., & Rinklebe, J., (2017). Interaction of arsenic with biochar in soil and water: A critical review. *Carbon, 113*, 219–230.

Williams, P. N., Price, A. H., Raab, A., Hossain, S. A., Feldmann, J., & Meharg, A. A., (2005). Variation in arsenic speciation and concentration in paddy rice related to dietary exposure. *Environmental Science & Technology, 39*(15), 5531–5540.

Wu, C., Huang, L., Xue, S. G., Pan, W. S., Zou, Q., Hartley, W., & Wong, M. H., (2017). Oxic and anoxic conditions affect arsenic (As) accumulation and arsinite transporter expression in rice. *Chemosphere, 168*, 969–975.

Xiang, Y., Xu, Z., Wei, Y., Zhou, Y., Yang, X., Yang, Y., & Zhou, Z., (2019). Carbon-based materials as adsorbent for antibiotics removal: Mechanisms and influencing factors. *Journal of Environmental Management, 237*, 128–138.

Yang, X., Iris, K. M., Tsang, D. C., Budarin, V. L., Clark, J. H., Wu, K. C. W., & Ok, Y. S., (2020). Ball-milled, solvent-free Sn-functionalization of wood waste biochar for sugar conversion in food waste valorization. *Journal of Cleaner Production, 268*, 122300.

Zhang, S., Geng, L., Fan, L., Zhang, M., Zhao, Q., Xue, P., & Liu, W., (2020). Spraying silicon to decrease inorganic arsenic accumulation in rice grain from arsenic-contaminated paddy soil. *Science of the Total Environment, 704*, 135239.

Zhang, X., Wang, H., He, L., Lu, K., Sarmah, A., Li, J., & Huang, H., (2013). Using biochar for remediation of soils contaminated with heavy metals and organic pollutants. *Environmental Science and Pollution Research, 20*(12), 8472–8483.

Zhao, F. J., McGrath, S. P., & Meharg, A. A., (2010). Arsenic as a food chain contaminant: Mechanisms of plant uptake and metabolism and mitigation strategies. *Annual Review of Plant Biology, 61*, 535–559.

Zheng, M. Z., Li, G., Sun, G. X., Shim, H., & Cai, C., (2013). Differential toxicity and accumulation of inorganic and methylated arsenic in rice. *Plant and Soil, 365*(1), 227–238.

Index

C

D

E

M

N

Q

R

T

U

V